T0176581

THE DATA INDUSTRY: THE BUSINESS AND ECONOMICS OF INFORMATION AND BIG DATA

THE DATA INDUSTRY: THE BUSINESS AND ECONOMICS OF INFORMATION AND BIG DATA

CHUNLEI TANG

Published by John Wiley & Sons, Inc., Hoboken, New JerseyPublished simultaneously in Canada

For general information on our other products and services or for technical support, please contact our Customer Care Department within the United States at (800) 762-2974, outside the United States at (317) 572-3993 or fax (317) 572-4002.

Wiley also publishes its books in a variety of electronic formats. Some content that appears in print may not be available in electronic formats. For more information about Wiley products, visit our web site at www.wiley.com.

Library of Congress Cataloging-in-Publication Data:

Names: Tang, Chunlei, author.
Title: The data industry : the business and economics of information and big
    data / Chunlei Tang.
Description: Hoboken, New Jersey : John Wiley & Sons, 2016. | Includes
    bibliographical references and index.
Identifiers: LCCN 2015044573 (print) | LCCN 2016006245 (ebook) | ISBN
    9781119138402 (cloth) | ISBN 9781119138419 (pdf) | ISBN 9781119138426
    (epub)
Subjects: LCSH: Information technology–Economic aspects. | Big
    data–Economic aspects.
Classification: LCC HC79.I55 T36 2016 (print) | LCC HC79.I55 (ebook) | DDC
    338.4/70057–dc23
LC record available at http://lccn.loc.gov/2015044573

Typeset in 10/12pt TimesLTStd by SPi Global, Chennai, India

Printed in the United States of America

10 9 8 7 6 5 4 3 2 1

BIBLIOGRAPHY

The data industry is a reversal, derivation, and upgrading of the information industry that touches nearly every aspect of modern life. This book is written to provide an introduction of this new industry to the field of economics. It is among the first books on this topic. The data industry ranges widely. Any domain (or field) can be called a "data industry" if it has a fundamental feature: the use of data technologies. This book (1) explains data resources; (2) introduces the data asset; (3) defines a data industry chain; (4) enumerates data enterprises' business models and operating model, as well as a mode of industrial development for the data industry; (5) describes five types of enterprise agglomeration, and multiple industrial cluster effects; and (6) provides a discussion on the establishment and development of data industry related laws and regulations.

DEDICATION

To my parents, for their tireless support and love
To my mentors, for their unquestioning support of my moving forward in my way

ENDORSEMENTS

"I have no doubt that data will become a fundamental resource, integrated into every fiber of our society. The data industry will produce incredible value in the future. Dr. Tang, a gifted young scientist in this field, gives a most up-to-date and systematic account of the fast-growing data industry. A must read of any practitioner in this area."

Chen, Yixin Ph.D.,
Full Professor of Computer Science and Engineering, Washington University in St Louis

"Data is a resource whose value can only be realized when analyzed effectively. Understanding what our data can tell us will help organizations lead successfully and accelerate business transformation." This book brings new insights into how to best optimize our learning from data, so critical to meeting the challenges of the future.

Volk, Lynn A., MHS,
Associate Director, Clinical and Quality Analysis, Information Services, Partners HealthCare

CONTENTS

PREFACE

In late 2009 my doctoral advisor, Dr. Yangyong Zhu at Fudan University published his book *Datalogy*, and sent me a copy as a gift. On the title page he wrote: "Every domain will be implicated in the development of data science theory and methodology, which definitely is becoming an emerging industry." For months I probed the meaning of these words before I felt able to discuss this point with him. As expected, he meant to encourage me to think deeply in this area and plan for a future career that combines my work experience and doctoral training in data science.

Ever since then, I have been thinking about this interdisciplinary problem. It took me a couple of years to collect my thoughts, and an additional year to write them down in the form of a book. I chose to put "data industry" in the book's title to impart the typical resource nature and technological feature of "data." That manuscript was published in Chinese in 2013 by Fudan University Press. In the title *The Data Industry*, I also wanted to clarify the essence of this new industry, which expands on the theory and concepts of data science, supports the frontier development of multiple scientific disciplines, and explains the natural correlation between data industrial clusters and present-day socioeconomic developments.

With the book now published, I intend to begin my journey into healthcare, with an ultimate goal of achieving the best in experience for all in healthcare through big data analytics. To date, healthcare has been a major battlefield of data innovations to help upgrade the collective human health experiences. In my postdoctoral research at Harvard, I work with Dr. David W. Bates, an internationally renowned expert on innovation science in healthcare. My focus is on commercialization-oriented healthcare services, and this has led to my engagement in several activities including composing materials of healthcare big data, proposing an Allergy Screener app, and designing a workout app for Promoting Bones Health in Children. However, there still exists a gap between data technology push and medical application pull. At present, many clinicians consider commercialization of healthcare data application to be irrelevant, and do not know how to translate research into technology commercialization, despite the fact that "big data" is at the peak of inflated expectations in Gartner's Hype Cycles. To address this gap, I plan to rewrite my book in English, mainly to address many of the shifting opinions, my own included.

Data science is an application-oriented technology as its developments are driven by the needs of other domains (e.g., financial, retail, manufacturing, medicine). Instead of replacing the specific area, data science serves as the foundation to improve and refine the performance of that area. There are two basic strengths of data technologies: one is its ability to promote the efficiency and increase the profit of existing industrial systems; the other is its application to identify hidden patterns and trends that cannot be found utilizing traditional analytic tools, human experience, or intuition. Findings concluded from data combined with human experience and rationality, are usually less influenced by prejudices. In my forthcoming book, I will discuss several scenarios on how to convert data-driven forces into productivities that can serve society.

Several colleagues have helped me in writing and revising this book, and have contributed to the formation of my viewpoints. I want to extend my special thanks to them for their valuable advice. Indeed, they are not just colleagues but dear friends Yajun Huang, Xiaojia Yu, Joseph M. Plasek, and Changzheng Yuan.

1

WHAT IS DATA INDUSTRY?

The next generation of information technology (IT) is an emerging and promising industry. But, what's truly the "next generation of IT"? Is it the next generation mobile networks (NGMN), Internet of Things (IoT), high-performance computing (HPC), or is it something else entirely? Opinions vary widely.

From the academic perspective, the debates, or arguments, over specific and sophisticated technical concepts are merely hype. How so? Let's take a quick look at the essence of information technology reform (IT reform) – digitization. Technically, it is a process that stores "information" that is generated in the real world from the human mind in digital form as "data" into cyberspace. No matter what types of new technologies emerge, the data will stay the same. As the British scholar Viktor Mayer-Schonberger once said [1], it's time to focus on the "I" in the IT reform. "I," as information, can only be obtained by analyzing data. The challenge we expect to face is the burst of a "data tsunami," or "data explosion," so data reform is already underway. The world of "being digital," as advocated some time ago by Nicholas Negroponte [2], has been gradually transformed to "being in cyberspace."[1]

With the "big data wave" touching nearly all human activities, not only are academic circles resolved to change the way of exploring the world as the "fourth paradigm"[2] but industrial community is looking forward to enjoying profits from

---

[1]Cyberspace, invented by the Canadian author William Gibson in his science fiction of *Neuromancer* (1984).

[2]The fourth paradigm was put forwarded by Jim Gray. http://research.microsoft.com/en-us/um/people/gray.

The Data Industry: The Business and Economics of Information and Big Data, First Edition. Chunlei Tang.
© 2016 John Wiley & Sons, Inc. Published 2016 by John Wiley & Sons, Inc.

"inexhaustible" data innovations. Admittedly, given the fact that the emerging data industry will form a strategic industry in the near future, this is not difficult to predict. So the initiative is ours to seize, and to encourage the enterprising individual who wants to seek means of creative destruction in a business startup or wants to revamp a traditional industry to secure its survival. We ask the reader to follow us, if only for a cursory glimpse into the emerging big data industry, which handily demonstrates the properties property of the four categories in Fisher–Clark's classification, which is to say: the resource property of primary industry, the manufacturing property of secondary industry, the service property of tertiary industry, and the "increasing profits of other industries" property of quaternary industry.

At present, industrial transformation and the emerging business of data industry are big challenges for most IT giants. Both the business magnate Warren Buffett and financial wizard George Soros are bullish that such transformations will happen. For example,[3] after IBM switched its business model to "big data," Buffett and Soros increased their holdings in IBM (2012) by 5.5 and 11%, respectively.

1.1 DATA

Scientists who are attempting to disclose the mysteries of humankind are usually interested in intelligence. For instance, Sir Francis Galton,[4] the founder of differential psychology, tried to evaluate human intelligence by measuring a subject's physical performance and sense perception. In 1971, another psychologist, Raymond Cattell, was acclaimed for establishing Crystallized Intelligence and Fluid Intelligence theories that differentiate general intelligence [3]. Crystallized Intelligence describes to "the ability to use skills, knowledge, and experience"[5] acquired by education and previous experiences, and this improves as a person ages. Fluid Intelligence is the biological capacity "to think logically and solve problems in novel situations, independently of acquired knowledge."[5]

The primary objective of twentieth-century IT reform was to endow the computing machine with "intelligence," "brainpower," and, in effect, "wisdom." This all started back in 1946 when John von Neumann, in supervising the manufacturing of the ENIAC (electronic numerical integrator and computer), observed several important differences between the functioning of the computer and the human mind (such as processing speed and parallelism) [4]. Like the human mind, the machine used a "storing device" to save data and a "binary system" to organize data. By this analogy, the complexities of machine's "memory" and "comprehension" could be worked out.

What, then, is data? Data is often regarded as the potential source of factual information or scientific knowledge, and data is physically stored in bytes (a unit of measurement). Data is a "discrete and objective" factual description related to an event,

---

[3]IBM's centenary: The test of time. The Economist. June 11, 2011. http://www.economist.com/node/18805483.
[4]https://en.wikipedia.org/wiki/Francis_Galton.
[5]http://en.wikipedia.org/wiki/Fluid_and_crystallized_intelligence.

and can consist of atomic data, data item, data object, and a data set, which is collected data [5]. Metadata, simply put, is data that describes data. Data that processes data, such as a program or software, is known as a data tool. A data set refers to a collection of data objects, a data object is defined in an assembly of data items, a data item can be seen as a quantity of atomic data, and an atomic data represents the lowest level of detail in all computer systems. A data item is used to describe the characteristics of data objects (naming and defining the data type) without an independent meaning. A data object can have other names [6] (record, point, vector, pattern, case, sample, observation, entity, etc.) based on a number of attributes (e.g., variable, feature, field, or dimension) by capturing what phenomena in nature.

1.1.1 Data Resources

Reaping the benefits of Moore's law, mass storage is generally credited for the drop in cost per megabyte from US$6,000 in 1955 to less than 1 cent in 2010, and the vast change in storage capacity makes big data storage feasible.

Moreover, today, data is being generated at a sharply growing speed. Even data that was handwritten several decades ago is collected and stored by new tools. To easily measure data size, the academic community has added terms that describe these new measurement units for storage: kilobyte (KB), megabyte (MB), gigabyte (GB), terabyte (TB), petabyte (PB), exabyte (EB), zettabyte (ZB), yottabyte (YB), nonabyte (NB), doggabyte (DB), and coydonbyte (CB).

To put this in perspective, we have, thanks to a special report, "All too much: monstrous amounts of data,"[6] in *The Economist* (in February 2010), an ingenious descriptions of the magnitude of these storage units. For instance, "a kilobyte can hold about half of a page of text, while a megabyte holds about 500 pages of text."[7] And on a larger scale, the data in the American Library of Congress amounts to 15 TB. Thus, if 1 ZB of 5 MB songs stored in MP3 format were played nonstop at the rate of 1 MB per minute, it would take 1.9 billion years to finish the playlist.

A study by Martin Hilbert of the University of Southern California and Priscila López of the Open University of Catalonia at Santiago provides another interesting observation: "the total amount of global data is 295 EB" [7]. A follow-up to this finding was done by the data storage giant EMC, which sponsored an "Explore the Digital Universe" market survey by the well-known organization IDC (International Data Corporation). Some subsequent surveys, from 2007 to 2011, were themed "The Diverse and Exploding Digital Universe," "The Expanding Digital Universe: A Forecast of Worldwide Information," "As the Economy Contracts, The Digital Universe Expands," "A Digital Universe – Are You Ready?" and "Extracting Value from Chaos."

The 2009 report estimated the scale of data for the year and pointed out that despite the Great Recession, total data increased by 62% compared to 2008, approaching 0.8 ZB. This report forecasted total data in 2010 to grow to 1.2 ZB. The 2010 report forecasted that total data in 2020 would be 44 times that of 2009, amounting to 35

---

[6]http://www.economist.com/node/15557421.
[7]http://www.wisegeek.org/how-much-text-is-in-a-kilobyte-or-megabyte.htm.

ZB. Additionally the increase in the amount of data objects would exceed that amount in total data. The 2011 report brought us further to the unsettling point that we have reached a stage where we need to look for a new data tool to handle the big data that is sure to change our lifestyles completely.

As data organizations connected by logics and data areas assembled by huge volumes of data reach a "certain scale," those massive different data sets become "data resources" [5]. The reason why a data resource can be one of the vital modern strategic resources for humans – even possibly exceeding, in the twenty-first century, the combined resources of oil, coal, and mineral products – is that currently all human activities, and without exception including the exploration, exploitation, transportation, processing, and sale of petroleum, coal, and mineral products, will generate and rely on data.

Today, data resources are generated and stored for many different scientific disciplines, such as astronomy, geography, geochemistry, geology, oceanography, aerograph, biology, and medical science. Moreover various large-scale transnational collaborative experiments continuously provide big data that can be captured, stored, communicated, aggregated, and analyzed, such as CERN's LHC (Large Hadron Collider),[8] American Pan-STARRS (Panoramic Survey Telescope and Rapid Response System),[9] Australian radio telescope SKA (Square Kilometre Array),[10] and INSDC (International Nucleotide Sequence Database Collaboration).[11] Additionally INSDC's mission is to capture, preserve, and present globally comprehensive public domain biological data. As for economic areas, there are the data resources constructed by financial organizations and the economic data, social behavior data, personal identity data, and Internet data, namely the data generated by social networking computations, electronic commerce, online games, emails, and instant messaging tools.

1.1.2 The Data Asset

As defined in academe, a standard asset has four characteristics: (1) it should have unexpired value, (2) it should be a debit balance, (3) it should be an economic resource, and (4) it should have future economic benefits. The US Financial Accounting Standards Board expands on this definition: "[assets are] probable future economic benefits obtained or controlled by a particular entity as a result of past transactions or events."[12] Basically, by this definition, assets have two properties: (1) an economic property, in that an asset must be able to produce an economic benefit, and (2) a legal property, in that an asset must be controllable.

Our now common understanding is that the intellectual asset, as one of the three key components[13] of intellectual capital, is a "special asset." This is based on the

---

[8]http://public.web.cern.ch/public/en/LHC/LHC-en.html.
[9]http://pan-starrs.ifa.hawaii.edu/public.
[10]http://www.ska.gov.au.
[11]http://www.insdc.org.
[12]http://accounting-financial-tax.com/2009/08/definition-of-assets-fasb-concept-statement-6.
[13]In the book *Value-Driven Intellectual Capital*, Sullivan argues that intellectual capital consists of intellectual assets, intellectual property and human assets.

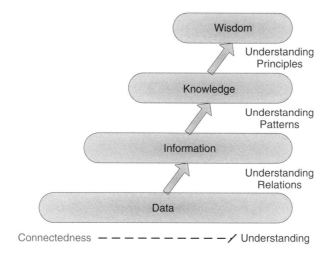

Figure 1.1 DIKW pyramid. Reproduced by permission of Gene Bellinger

concept of intellectual capital introduced in 1969 by John Galbraith, an institutional economist of the Keynesian school, and later expanded by deductive argument due to Annie Brooking [8], Thomas Stewart [9], and Patrick Sullivan [10]. In more recent years the concept of intellectual asset was further refined to a stepwise process by the British business theorist Max Boisot, who theorized on the "knowledge asset" (1999) [11]; by Chicago School of Economics George Stigler, who added an "information asset" (2003) [12]; and by DataFlux CEO Tony Fisher, who suggested a "data asset" specification process (2009) [13] that would closely follow the rules presented in the DIKW (data, information, knowledge, and wisdom) pyramid shown in Figure 1.1.

According to the ISO 27001:2005 standard, data assets are an important component of information assets, in that they contain source code, applications, development tools, operational software, database information, technical proposals and reports, job records, configuration files, topological graphs, system message lists, and statistical data.

We therefore want to treat *data asset* in the broadest sense of the term. That is to say, we want to redefine the data asset as data exceeding a certain scale that is owned or controlled by a specific agent, collected from the agent's past transactions involved in information processes, and capable of bringing future economic benefits to the agent.

According to Fisher's book *The Data Asset*, the administrative capacity of a data asset may decide competitive advantages of an individual enterprise, so as to mitigate risk, control cost, optimize revenue, and increase business capacity, as is shown in Figure 1.2. In other words, the data asset management perspective should closely follow the data throughout its life cycle, from discovery, design, delivery, support, to archive.

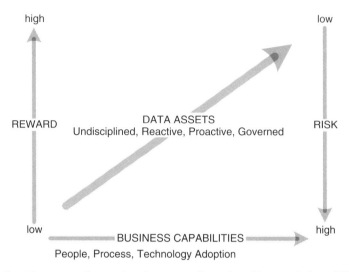

Figure 1.2 Advantages of managing data assets. Reproduced by permission of Wiley [13]

Our view[14] is that the primary value of data assets lies in the willingness of people to use data, and for some purpose as is reflected by human activities arising from data ownership or application of data. In a sense, data ownership, which defines and provides information about the rightful owner of data assets, depends on the "granularity of data items." Here is a brief clinical example of how to determine data ownership. Diagnostic records are associated with (1) patient's disease status, in terms of disease activity, disease progression, and prognosis, and (2) physician's medical experience with symptoms, diagnosis, and treatments. Strictly speaking, the patient and physician are both data owners of diagnostic records. However, we can minimize diagnostic records to patient's disease status, namely reduce its granularity such that only the patient takes data ownership of the diagnostic records.

1.2 INDUSTRY

The division of labor mentioned in one of Adam Smith's two classic works *An Inquiry into the Nature and Causes of the Wealth of Nations* (1776), is generally recognized as the foundation of industry [14], the industry cluster, and other industry schemes.

Industry is the inevitable outcome of the social division of labor. It was spawned by scientific and technological progress and by the market economy. Industry is in fact a generic term for a market composed of various businesses having interrelated benefits and related divisions of labor.

---

[14]This view is based on a discussion my peers and I had with Dr. Yike Guo, who is the founding director at Data Science Institute as well as Professor of Computer Science, Imperial College London.

1.2.1 Industry Classification

In economics, classification is usually the starting point and the foundation of research for industries. Industries can be classified in various ways:

- *By Economic Activity.* Primary industry refers to all the resource industries dealing with "the extraction of resources directly from the Earth," secondary industry to industries involved in "the processing products from primary industries," tertiary industry to all service industries, and the quaternary industry to industries that can significantly increase the industrial profits of other industries. The classification of tertiary industries is due to Fisher (1935) and the classification of quaternary industries is due to Clark (1940).
- *By Level of Industrial Activity.* There are three levels: use of similar products as differentiated by an "industrial organization," use of similar technologies or processes as differentiated by an "industrial linkage," and use of similar economic activities as differentiated by an "industrial structure."
- *By a System of Standards.* For international classification standards, we have the North American Industry Classification System (NAICS), International Standard Industrial Classification of All Economic Activities (ISIC), and so forth.

Of course, industries can be further identified by products, such as the chemical industry, petroleum industry, automotive industry, electronic industry, meatpacking industry, hospitality industry, food industry, fish industry, software industry, paper industry, entertainment industry, and semiconductor industry.

1.2.2 The Modern Industrial System

Computational optimization, modeling, and simulation as a paradigm not only produced IT reform of the information industry but also a fuzzy technology border, as new trends were added to the industry, such as software as a service, embedded software, and integrated networks. In this way, IT reform atomized the traditional industries and transformed their operation modes, thus prompting the birth of a new industrial system. The industries in this modern industrial system include, but are not limited to, the knowledge economy, high-technology industry, information industry, creative industries, cultural industries, and wisdom industry.

Knowledge Economy The "knowledge economy" is a term introduced by Austrian economist Fritz Machlup of Princeton University in his book *The Production and Distribution of Knowledge in the United States* (1962). It is a general category that has enabled the classification of education, research and development (R&D), and information service industries, but excluding "knowledge-intensive manufacturing," in "an economy directly based on the production, distribution, and use of knowledge and information," in accord with the 1997 definition by the OECD (Organization for Economic Co-operation and Development).

High-Technology Industry The high-technology industry is a derivative of the knowledge economy that uses "R&D intensity" and "percentage of R&D employees" as a standard of classification. The main fields are information, biology, new materials, aerospace, nuclear, and ocean, and characterized by (1) high demand for scientific research and intensity of R&D expenditure, (2) high level of innovativeness, (3) fast diffusion of technological innovations, (4) fast process of obsolescence of the prepared products and technologies, (5) high level of employment of scientific and technical personnel, (6) high capital expenditure and high rotation level of technical equipment, (7) high investment risk and fast process of the investment devaluation, (8) intense strategic domestic and international cooperation with other high-technology enterprises and scientific and research centers, (9) implication of technical knowledge in the form of numerous patents and licenses, (10) increasing competition in international trade.

Information Industry The "information industry" concept was developed in the 1970s and is also associated with the pioneering efforts of Machlup. In 1977 it was advanced by Marc Uri Porat [15] who estimated the predominant occupational sector in 1960 was involved in information work, and established Porat's measurements. The North American Industry Classification System (NAICS) sanctioned the information industry as an independent sector in 1997. According to the NAICS, the information industry includes three establishments engaged "(1) producing and distributing information and cultural products," "(2) providing the means to transmit or distribute these products as well as data or communications," and "(3) processing data."

Creative Industries Paul Romer, an endogenous growth theorist, suggested in 1986 that countless derived new products, new markets, and new opportunities for wealth creation [16] could lead to the creation of new industries. Although Australia put forward in 1994 the concept of a "creative nation," Britain was first to actually give us a manifestation of the "creative industries" when it established a new strategic industry with the support of national policy. According to the UK Creative Industries Mapping Document (DCMS) definition, creative industries as an industry whose "origin (is) in individual creativity, skill and talent and which has a potential for wealth and job creation through the generation and exploitation of intellectual property (1998)." This concept right away swept the globe. From London it spread to New York, Tokyo, Paris, Singapore, Beijing, Shanghai, and Hong Kong.

Cultural Industries The notion of a culture industry can be credited to the popularity of mass culture. The term "cultural industries" was coined by the critical theorists Max Horkheimer and Theodor Adorno. In the post-industrial age, overproduction of material similarly influenced culture, to the extent that the monopoly of traditional personal creations was broken. To criticize such "logic of domination in post-enlightenment modern society by monopoly capitalism or the nation state," Horkheimer and Adorno argued that "in attempting to realise enlightenment values of reason and order, the holistic power of the individual is undermined."[15] Walter Benjamin, an eclectic thinker also from the Frankfurt School, had the opposite view.

---

[15]http://en.wikipedia.org/wiki/Culture_industry.

He regarded culture as due to "technological advancements in art." The divergence of those views reflects the process of culture "from elites to the common people" or "from religious to secular," and it is such argumentations that accelerated culture industrialization to emerge as the "cultural industry." In the 1960s, the Council of Europe and UNESCO (United Nations Educational, Scientific and Cultural Organization) changed "industry" to the plural form "industries," to effect a type of industry economy in a broader sense. In 1993, the UNESCO revised the 1986 cultural statistics framework, and defined the cultural industries as "those industries which produce tangible or intangible artistic and creative outputs, and which have a potential for wealth creation and income generation through the exploitation of cultural assets and production of knowledge-based goods and services (both traditional and contemporary)." Additionally what cultural industries "have in common is that they all use creativity, cultural knowledge, and intellectual property to produce products and services with social and cultural meaning." The cultural industries therefore include cultural heritage, publishing and printing, literature, music, performance art, visual arts, new digital media, sociocultural activities, sports and games, environment, and nature.

Wisdom Industry   Taking the lead in exalting "wisdom," in a commercial sense, IBM has been a vital player in the building of a "Smarter Planet" (2008). In the past IBM had advanced two other such commercial hypes: "e-Business" in 1996 and "e-Business on Demand" in 2002. These commercial concepts, as they were expanded both in connotation and denotation, allowed IBM to thus explore both market depth and width. With the intensive propaganda related to Cloud computing and the IoT, there are now hundreds of Chinese second-tier and third-tier cities that have discussed constructing a "Smart City." In the last couple of years IBM has won bids for huge projects in Shenyang, Nanjing, Shenzhen, among other places. To the best of our knowledge, however, the wisdom industry, which has only temporarily appeared in China, is based on machines and, we believe, will never have the ability to possess wisdom, knowledge, and even information, without the human input of data and thus data mining.

From these related descriptions of industries, we can see that cultural industries have a relatively broad interpretation. The United States treats cultural industries as copyright industries in the commercial and legal sense, whereas Japan has shifted to the expression "content industries" based on the transmission medium. In the inclination to emphasize "intellectual property" over "commoditization," the wisdom industry, knowledge economy, and information industry (disregarding the present order of appearance) are externally in compliance with the DIKW pyramid. The information industry may be further divided into two sectors. The first sector is the hardware manufacturing sector that includes equipment manufacturing, optical communication, mobile communication, integrated circuit, display device, and application electronics. The second is the information component of the services sector that includes the software industry, network information service (NIS), digital publishing, interactive entertainment, and telecommunications service.[16] The wisdom industry, which is essentially commercial hype despite being labeled "an upgraded version of creative

[16]https://en.wikipedia.org/wiki/Telecommunications_service.

industries," is no more than a use of "human beings" disguised as industrial carriers to "machines."

1.3   DATA INDUSTRY

From the foregoing description one could say that the information industry may be simply understood as digitization. Technically, IT is a process that stores "information" generated in the real world by human minds in digital form amassed as "data" in cyberspace, as is the process of producing data. In time the accumulated data can be sourced from multiple domains and distinct sectors.

The mining of "data resources" and extracting useful information already is seemingly "inexhaustible" as data innovations keep on emerging. Thus, to effectively endow all the data innovations with a business model – namely industrialization – would call for us to rename this strategic emerging industry, which is strong enough to influence the world economy, "data industry." The data industry is the reversal, derivation, and upgrading of the information industry.

1.3.1   Definitions

Connotation and denotation are two principal ways of describing objects, events, or relationships. Connotation relates to a wide variety of natural associations, whereas denotation consists in a precise description. Here, based on these two types of descriptions, we offer two definitions, in both a wide and a narrow sense, for the data industry.

In a wide sense, the data industry has evolved three technical processes: data preparation, data mining, and visualization. By these means, the data industry connotes rational development and utilization of data resources, effective management of data assets, breakthrough innovation of data technologies, and direct commoditization of data products. Accordingly, by definition then, the existing industrial sectors – such as publishing and printing, new digital media, electronic library and intelligence, digital content, specific domain data resources development, and data services in distinct sectors – should be included in the data industry. To these we should add the existing data innovations of web creations, data marketing, push services, price comparison, and disease prevention.

In a narrow sense, the data industry is usually divided into three major components: upstream, midstream, and downstream. In this regard, by definition, the data industry denotes data acquisition, data storage, data management, data processing, data mining, data analysis,[17] data presentation, data product pricing, valuation, and trading.

1.3.2   An Industry Structure Study

To understand profitability of a new industry, one must look at the distinctive structure that shapes the unfolding nature of competitive interactions. On the surface, the data industry is extremely complex. However, there are only four connotative factors associated with the data industry. These factors include: data resources, data assets,

---

[17]In this book, from the perspective of *Data Science*, I try to distinguish data mining and traditional data analysis tools or techniques, the latter refer to data analysis.

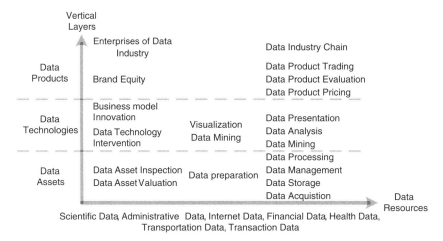

Figure 1.3   Structure of the data industry

data technologies, and data products. In a nutshell, from a vertical bottom-top view, the structure of the data industry (as shown in Figure 1.3) could be expressed by (1) data assets precipitation that forms the foundation of the data industry, (2) data technologies innovation as its core, and (3) data products circulation as its means. Theoretically, these three layers rely on data sources via mutually independent units that form underlying substructures, and then vertically form the entire data industry chain.

Technology Substructure   The essence of the industry is to cope with conversion technologies. The corresponding term for the data industry is "data science," which is "a continuation of some of the data analysis fields such as statistics, data mining, and predictive analytics, similar to Knowledge Discovery in Databases (KDD)."[18]

Peter Naur, a Danish pioneer in computer science and Turing award winner, once coined a new word – dataloy – in 1966 because he disliked the term computer science. Subsequently datalogy was adopted in Denmark and in Sweden as datalogi. However, Naur lost to William Cleveland of Purdue University in the influence of new word combinations or coinages, despite the fact that Naur was far more known than Cleveland. In 2001, Cleveland suggested a new word combination – data science – as an extension of statistics and, using that term, published two academic journals *Data Science Journal* and *The Journal of Data Science* (on the two disciplines of statistics) in 2002 and 2003, respectively. Cleveland's proposal has had an enormous impact over the years. Whenever or wherever people mentioned "data analysis" now, they first associate it with statistical models. Yet this curious episode did not stop data technologies from evolving.

Back to the technology substructure, the data industry has developed through the following three steps.

Step 1: *Data Preparation.*   Similar to the geological survey and analysis during mineral exploration [5], data preparation determines data quality and

---

[18]https://en.wikipedia.org/wiki/Data_science.

selection of follow-up mining. These methods include (1) judgment of the availability of a data set (e.g., if a data source is isomeric and the data set is accessible); (2) analysis of the physical and logical structure of a data set; and (3) metadata acquisition and integration.

Step 2: *Data Mining.*   As an efficient and scalable tool, data mining draws on ideas [6] from other disciplines. The ideas include (1) query optimization techniques, like indexing, labeling, and join algorithms, to enhance query processing from traditional database technologies; (2) sampling, estimation, and hypothesis testing from statistics; (3) search algorithms, modeling techniques, and learning theories from artificial intelligence, pattern recognition, and machine learning; and (4) high-performance or parallel computing, optimization, evolutionary computing, information theory, signal processing, and information retrieval from other areas. In general, data mining tasks are divided into two categories [6]: predictive tasks and descriptive tasks. Both of these data mining processes utilize massive volumes of data and exploratory rules to discover hidden patterns or trends in the data that cannot be found with traditional analytical tools or by human intuition.

Step 3: *Visualization.*   The idea of visualization originated with images created by computer graphics. Exploration in the field of information visualization [17] became popular in the early 1990s, and was used to help understand abstract analytic results. Visualization has remained an effective way to illuminate cognitively demanding tasks. Cognitive applications increased in sync with the large heterogeneous data sets in fields such as retail, finance, management, and digital media. Data visualization [18], an emerging word combination containing both "scientific visualization" and "information visualization," has been gradually accepted. Its scope has been extended to include the interpretation of data through 3D graphics modeling, image rendering, and animation expression.

Resource Substructure   Data resources have problems similar to those of traditional climate, land, and mineral resources. These include an uneven distribution of resource endowments, reverse configuration of production and use, and difficulties in development. That is to say, a single property or combination of properties of a data resource (e.g., diversity, high dimensionality, complexity, and uncertainty) can simultaneously reflect the position and degree of priority for a specific region within a given time frame so as to directly dictate regional market performance.

The resource substructure of the data industry consists of (1) a resource spatial structure (i.e., the spatial distribution of isomorphic data resources in different regions); (2) a resource type structure (i.e., the spatial distribution of non-isomorphic data resources in the same region); (3) a resource development structure (i.e., the spatial-temporal distribution of either to-be-developed data resources or having-been-developed data resources that were allowed for development); (4) a resource utilization structure (i.e., the spatial-temporal distribution of multilevel deep processing of having-been-developed data resource); and (5) a resource protection structure (i.e., the spatial-temporal distribution of protected data resources according to a specific demand or a particular purpose).

Sector Substructure   The sector substructure of the data industry is based on the relationships of various data products arising from the commonness and individuality in the processing of production, circulation, distribution, and consumption.

In regard to the information industry, sub-industries of the data industry may have two methods of division. First is whether data products are produced. This can be divided into (a) nonproductive sub-industry and (b) productive sub-industry. In this regard data acquisition, data storage, and data management belong to the nonproductive sub-industry, and in the productive sub-industry, data processing and data visualization directly produce data products while data pricing, valuation, and trading indirectly produce data products. Second is whether data products are available to a society. Data product availability can be divided into (c) an output projection sub-industry and (d) an inner circulation sub-industry, whereby the former provides data products directly to society and the latter provides data products within a sub-industry or to other sub-industries.

1.3.3   Industrial Behavior

Industrial behavior of the data industry is concentrated on four areas: data scientist (or quant [19]), data privacy, product pricing, and product rivalry.

Data Scientist   Victor Fuchs, often called the "Dean of health economists", named the physician "the captain of the team" in his book *Who Shall Live? Health, Economics, and Social Choice* (1974). Data scientists could be similarly regarded the "captains" of the data industry.

In October 2010, *Harvard Business Review* announced[19] that the data scientist has been becoming "the sexiest job of the 21st century." Let's look at what it means to be called "sexiest." It is not only the attraction of this career path that is implied, it is more likely the art implied by "having rare qualities that are much in demand." The authors of this HBS article were Thomas Davenport and D. J. Patil, both men well known in academe and in industrial circles. Davenport is a famous academic author, and the former chief of the Accenture Institute for Strategic Change (now called Accenture Institute for High Performance Business, based in Cambridge, Massachusetts). Davenport was named one of the world's "Top 25 Consultants" by *Consulting* in 2003. Patil is copartner at Greylock Partners, and was named the first US Chief Data Scientist by the White House in February 2015. In the article they described the data scientist as a person having clear data insights through the use of scientific methods and mining tools. Data scientists need to test hunches, find patterns, and form theories. Data scientists not only need to have a professional background in "math, statistics, probability, or computer science" but must also have "a feel for business issues and empathy for customers." In particular, the top data scientists should be developers of new data mining algorithms or innovators of data products and/or processes.

---

[19]https://hbr.org/2012/10/data-scientist-the-sexiest-job-of-the-21st-century.

According to an earlier report by the McKinsey Global Institute,[20] data scientists are in demand worldwide and their talents are especially highly sought after by many large corporations like Google, Facebook, StumbleUpon, and Paypal. Almost 80% of the related employees think that the yearly salary of this profession is expected to rise. The yearly salary for a vice president of operations may be as high as US$132,000. MGI estimated that "by 2018, in the United States, 4 million positions will require skills" gained from experience working with big data and "there is a potential shortfall of 1.5 million data-savvy managers and analysts."

Data Privacy Russian-American philosopher Ayn Rand wrote in his 1943 book *The Fountainhead* that "Civilization is the progress toward a society of privacy." As social activities increasingly "go digital," privacy becomes more of an issue related to posted data. Every January 28 is designated as Data Privacy Day (DPD) in the United States, Canada, and 47 European countries, to "raise awareness and promote privacy and data protection best practices."[21]

Private data includes medical and social insurance records, traffic tickets, credit history, and other financial information. There is a striking metaphor on the Internet: computers, laptops, and smart phones are the "windows" – that is to say, more and more people (not just identifying thieves and fraudsters) are trying to break them into your "private home," to access your private information. The simple logic behind this metaphor is that your private data, if available in sufficient quantity for analysis, can have huge commercial interest for some people.

Over the past several years, much attention has been paid to private data snooping, and to the storage of tremendous amounts of raw data in the name of national security. For instance, in 2011, Google received 12,271 requests to hand over its users' private data to US government agencies, and among them law enforcement agencies, according to company's annual Transparency Report. Telecom operators responded to "a portion of the 1.3 million"[22] law enforcement requests for text messages and phone location data were largely without issued warrants. However, a much greater and more immediate data privacy threat is coming from large number of companies, probably never even heard of, called "data brokers."[23] They are electronically collecting, analyzing, and packaging some of the most sensitive personal information and often electronically selling it without the owner's direct knowledge to other companies, advertisers, and even the government as a commodity. A larger data broker named Acxiom, for example, has boasted that it has, on average, "1,500 pieces of information on more than 200 million Americans [as of 2014]."[23]

No doubt, data privacy will be a central issue for many years to come. The right of transfer options for private electronic data should be returned to owners from the handful of companies that profiteer by utilizing other people's private information.

Product Pricing We use the search engine (a primary data product) to demonstrate how to price a product. It is noteworthy that a search engine is not really software and

---

[20]http://www.mckinsey.com/insights/business_technology/big_data_the_next_frontier_for_innovation.
[21]http://en.wikipedia.org/wiki/Data_Privacy_Day.
[22]http://www.wired.com/2012/07/massive-phone-surveillance.
[23]http://www.cbsnews.com/news/the-data-brokers-selling-your-personal-information.

is not really free. As early as 1998, Bill Gross, the founder of GoTo.com, Inc. (now called Overture), applied for a patent for search engine pricing.

Today's popular search engines operate using an open and free business model, meaning they do not make money from users but instead are paid by advertisers. There are two types of advertising in the search engine. One is the pay-per-click (PPC) model used by Google, whereby no payment is solicited from the advertiser if no user clicks on the ad. The other is the "ranking bid" model "innovated" by Baidu, whereby search results are ranked according to the payment made by advertisers. Google, in October 2010, adjusted its cost-per-click (CPC) pricing by adding a 49% premium to wrestler-type advertising sponsors[24] who want to take the optimum position in the results. Cost-per-click is similar to Baidu's left ranking that has existed for long time and contributes almost 80% of the revenue from advertisements.

Compared to the search engine, targeted advertising is a more advanced data product. Targeted advertising consists of community marketing, mobile marketing, effect marketing, interaction innovation, search engine optimization (SEO), and advertisement effect monitoring. Despite the fact that targeted advertising "pushes" goods information to the consumers, its vital function is to "pull," to exploit the vicissitudes and chaotic behavior of consumers. One way is by classifying users through the tracking and mining of Cookie files in users' browsers and then associating these classes by matching related products along with sponsor rankings. Another way is to monitor users' mouse movements by calculating residence time to try and determine the pros and cons of an interactive pop-up ad. Yet, there are far more than these two ways to target consumers, such as by listening to background noise (music, wind, breathing, etc.) produced by a user's laptop microphone. In sum, the purpose of targeted advertising is to nudge customer interest preferences to the operational level, with plenty of buying options to increase revenue, enhance the interactive experience, retain customer loyalty, and reduce the cost of user recall.

Product Rivalry

Oligarch Constraint: e-Books   Here we only focus on the content of e-Books in digital form, without their carriers – computers, tablets, smart phones, and other electronic devices.

When Richard Blumenthal, the senior US Senator from Connecticut, served as Attorney General of Connecticut, he sent a letter of inquiry in August 2010 to Amazon regarding antitrust scrutiny on the pricing of e-Books. Blumenthal, undoubtedly, thought that the accord between sellers and publishers on e-Book pricing was bound to the increase chance of monopoly pricing, in "driving down prices in stock and pushing up prices in sales" adopted by Amazon to suppress smaller competitors. Today, there are over 3.5 million e-Books available in the Kindle Store of Amazon, and most of them are sold for less than US$10.

---

[24]Wrestler-type advertising sponsors refer to those who are willing to pay high to advertise their poor-quality products.

Weakening of Oligarch Restriction: eBooks Searched via Price Comparison In December 2010, Google entered this chaotic e-Book market, and claimed an "all about choice" strategy, which is to say, (1) any devices, including Android and iOS devices, browsers, special eBook readers (e.g., Amazon's Kindle, Barnes & Noble's Nook); (2) any book Google would ultimately provide, amounting to more than 130 million e-Books worldwide, with an initial 3 million volumes online including scanned edition of unique copies;[25] (3) any payment options, e-Books that could be bought from Google's Checkout using various payment options. Interestingly, Google added a function allowing price comparison of its e-Books to thousands of cooperative retailers. James McQuivey, a vice president and principal analyst at Forrester Research, commented that Google opened a gate to about 4,000 retailers who previously did not have the capability to invest in large-scale technology necessary to surmount powerful market competition.

1.3.4 Market Performance

Relative to this text three metric standards are used for measuring the market performance of the data industry: product differentiation, efficiency and productivity, and competition.

Product Differentiation In traditional industries, despite excluding a purely competitive market as well as an oligopoly market [20], product variety is considered to differ by some degree of innovation to allow for product differentiation (or simply differentiation). In other words, launching a new product is better than changing the packaging, advertising theme, or functional features of a product. In the data industry, however, we cannot say that product variety is more relevant to innovation compared to product differentiation. For example, for search engines, a plurality of search engines (e.g., keyword search tools, image search engines) belong to product differentiation, but they are based on different data technologies and depend on supply, demand, and consumers of various walks of life, in communities and organizations. Particularly, data product differentiation is controlled by the scale and diversity of data resources, such that even using the same algorithm in different domains will result in different data products.

Efficiency and Productivity Efficiency is "the extent to which time, effort, or cost is used well for the intended task or function,"[26] and this is usually classified into technical efficiency (and technological advance where the time factor is considered), cost efficiency, allocative efficiency, and scale efficiency. Productivity is an efficiency of production activities [21], and this can be expressed as a function, namely the ratio of output to inputs used in the production process. When the production process involves a single input and a single output, the production function can

---

[25]These so-called unique copies originated from the controversial Google Books Library Project launched in 2004 under the assistance of five partners: Harvard University Library, Stanford University Library, Oxford University Library, Michigan University Library, and New York Public Library.
[26]http://en.wikipedia.org/wiki/Efficiency.

be used to indicate productivity. When integrating multiple inputs or multiple outputs, total factor productivity is needed to show a change (increase or decrease) in productivity.

In 2001, after researching the efficiency and productivity of the information industry between 1995 and 1999, economist Dale Jorgenson of Harvard [22] pointed out that, over 50% of the entire technological advance in the US economy should be attributed to the technological advances in information hardware manufacturing. China and Japan also witnessed fast advancement in this sector. However, recent research shows[27] that from 2002 to 2006, the total factor productivity of China's software industry was 3.1% while the gain in technical efficiency was only 0.9%. This shows clearly that expensive hardware replacement and slow software innovations can no longer rapidly push economic growth. In addition, users have begun customizing data products according to their own demands, instead of buying standards-based servers, software, and solutions.

Competition Unlike other industries, competition in the data industry covers political, economic, military, and cultural areas – from the microscopic to the macroscopic and from virtual to real. Big data has already encroached on such fields, directly affecting our lives, as aerospace, aviation, energy, electric power, transportation, healthcare, and education. But the data industry faces competition both within a nation's borders and beyond its borders, which is to say, international competition. In the future it is probable that this international data competition will cause nations to compete for digital sovereignty in accord with the scale and activity of the data owned by a country and its capability of the interpreting and utilizing data. Cyberspace may prove to be another gaming arena for great powers, besides the usual border, coastal, and air defense tactics.

In the United States, in 2003, the White House published *The National Strategy to Secure Cyberspace*, a document that defines the security of cyberspace as a subset of Homeland Security. The US Air Force (USAF) answered that call in December 2005 when it enlarged the scope of its operational mission to fly and fight in air, space, and cyberspace. One year later, during a media conference, the USAF announced the establishment of an Air Force Cyberspace Command (USCYBERCOM). In March 2008, the USCYBERCOM released its strategic plan, and set new requirements for the traditional three missions of the USAF. These include (1) global vigilance: perception and transfer; (2) global reach: connection and transmission; and (3) global power: determent and crackdown. In 2009, President Obama personally took charge of a cyberspace R&D project where the core content is data resource acquisition, integration and processing, and utilization. In the same year, Obama issued a presidential national security order that set the cybersecurity policy as a national policy priority, and defined cyberspace crime as unauthorized entry and acquisition of data. In September 2010, the US military forces successfully destroyed the nuclear facilities of Iran through the virus "stwxnet" that was hidden in a flash driver, starting a war in

[27] Source: Li, He. A Study of Total Factor Productivity of Software Industry in China. Master's Thesis of Zhejiang Technology and Business University, 2008.

cyberspace. On March 29, 2012, the Obama administration released the *Big Data Research and Development Initiative*, which included the Department of Defense, Defense Advanced Research Projects Agency, National Science Foundation, National Institutes of Health, Department of Energy, and US Geological Survey. The six federal departments and agencies made commitments to invest over 200 million dollars altogether, "to greatly improve the tools and techniques needed to access, organize, and glean discoveries from huge volumes of digital data." In another related development,[28] the Pentagon has approved a major expansion of the USCYBERCOM in January 2013 over several years, "increasing its size more than fivefold" – "Cyber command, made up of about 900 personnel, will expand to include 4,900 troops and civilians." Recently, the USCYBERCOM appears to be more urgent the need to reach "a goal of 6,000 person"[29] by the end of 2016.

So far as the United States is concerned, it has implemented an entire force operational roadmap of both internal and external cyberspace as well as data resource protection, utilization, and development. By now, Russia, Britain, Germany, India, Korea, and Japan are doing similar work.

It should be noted that unlike previous warfare, this so-called sixth-generation war[30] is showing much heavier dependence on the industry sector. For example, when the United States carried out its cyberspace maneuvers, the participants included multiple government departments and related private sector companies, in addition to the operational units. Future international data industry competitions will ultimately shape the competitive advantages of all countries in cyberspace.

---

[28] http://www.washingtonpost.com/world/national-security/pentagon-to-boost-cybersecurity-force/ 2013/01/27/d87d9dc2-5fec-11e2-b05a-605528f6b712_story.html.
[29] http://www.defenseone.com/threats/2015/02/us-cyber-command-has-just-half-staff-it-needs/104847.
[30] Russia refers the cyber warfare as the "sixth-generation war."

2

DATA RESOURCES

Data and a data resource can, respectively, be understood as data that has a solitary existence in cyberspace and as a data resource that is collected and stored with similar data to a certain scale. This kind of relationship is analogous to gathering some phone numbers in a city – the data is a single resident's phone number, and a data resource refers to phone numbers of all residents in a city.

The standard way to classify data resources is as general and dedicated data resources according to data organization, and further by how accessible the data is, as judged by whether the data resource is sensitive or is available for public use. The general data resources refer to the database systems that assembled data during the early stages of information processing (Oracle, SQL Server, DB2, etc.), and the dedicated data resources to geographic data, medical images (X-ray film, MRI and CT scans, etc.), and multimedia whose processing is by dedicated equipment or software. In this regard the term "sensitive" is purely descriptive; whether a data resource is sensitive or publicly available must fit criteria guided by law.

In this chapter, we discuss the typical applications of data resources in data science and classify data resources into seven types under various specific domains.

2.1 SCIENTIFIC DATA

Data is currently a major study objective in scientific research. Data science has already been formed as a discipline, and is now supporting all research works related to both natural and social sciences. Not so long ago, this was novel idea, but today

The Data Industry: The Business and Economics of Information and Big Data, First Edition. Chunlei Tang.
© 2016 John Wiley & Sons, Inc. Published 2016 by John Wiley & Sons, Inc.

indeed, just like Bill Howe[1] of the University of Washington recently observed, "All science is fast becoming what is called data science."

2.1.1 Data-Intensive Discovery in the Natural Sciences

Turing Award winner (1998) Jim Gray once made a prophetic observation [23], "We are seeing the evolution of two branches of every discipline." Anyone today who conducts a research study will readily agree that this is indeed the case. For example, if we compare bioinformatics and computational biology, the informaticist collects and analyzes data from many different experiments, while the computational biologist simulates the working process of the biological systems to learn how the protein or cell metabolites or behaves.

In the natural sciences, demand-driven research has shattered the age-old "question and answer" paradigm. This trend today is to focus on (1) the process of creation, meaning action is taken based on external driving forces instead of the best acquired knowledge, such as social usefulness and commercial value; (2) the ability to finalize results, meaning pursue usefulness instead of completion or perfection, such as a new method that effectively utilizes incomplete and imperfect knowledge; (3) integration of disciplines, meaning on extensibility and robustness, such as using in the analysis, or explanation, knowledge from other disciplines.

Take the Earth and environmental sciences, for example. This discipline evolved from the combined disciplines of geology, atmospheric chemistry, ecosystem, and Earth system. Then, to tackle current environmental application problems, it was coupled with the combined disciplines of atmospheric science, astronomy, biology, and praxeology. Consider the water supplied by snowmelt runoff [23]. Worldwide, about 1 billion people depend on snowmelt or glacier melt for the purpose of drinking, irrigation, power generation, and entertainment. Traditional water resources management (WRM) is not applicable under subsistence conditions of a population swell due to relocation, climate change, and land use. The scientific community has delved into historical data to design solutions to this societal need in order to learn more about the fundamental process-based water cycle. Critical to estimating the spatial distribution and heterogeneity of the snow water equivalent turned out to be satellite data on the high reflectivity of snow cover, and remote measurement data of snow-to-soil humidity, to make up for the lack of knowledge about "snowfall in watershed and water balance" and the "rain-on-snow runoff." Thus WRM researchers were able to get better knowledge of the timing and magnitude of runoffs.

In addition, data experimentation with data [5] began to be accepted and recognized. It is characteristic that the experimental objects and tools of such a research approach are both composed of data: data verifies data through data. In July 2012, the CERN announced the successful finding of the suspected "God Particle" – the Higgs Boson, by relying on two independent data experiments, the Atlas and CMS. The data used in the experiments was then acquired by the Large Hadron Collider (LHC).

---

[1]Quoted from Bill Howe's tweet dated May 15, 2013: "All science is fast becoming what is called data science."

2.1.2   The Social Sciences Revolution

Duncan Watts [24], who proved the theory of six degrees of separation of human interconnectedness and then proposed the small world theory, also proffered that "if handled appropriately, data about Internet-based communication and interactivity could revolutionize our understanding of collective human behavior" and make social science to be considered "the science of the twenty-first century."

In the social sciences, such a change is occurring: (1) In the way data are acquired; data on social behavior is gradually replacing the traditional survey for research on dynamic interpersonal communication, as large social networks evolve into interactive online communities, influencing public opinion through Internet Cookies and massive-scale emotional contagion studies of social networks. (2) In the way data are processed; data mining may replace the third paradigm of hypothesize and test. Take efficiency and productivity, for example, some current methods utilizing least squares estimation of econometric frontier models, total factor productivity, data development analysis, and stochastic frontier analysis have very strict data requirements, and sometimes even directly determine the data. (3) In the way data are quantified; all available data may replace samples using stochastic sampling, due to the inherent flaws of ductility and unsuitability for investigating subclasses.

In 2012, President Obama broke the spell that "no president since the Great Depression has won reelection with a jobless rate higher than 7.4%." The secret, uncovered by *TIME*'s White House correspondent Michael Scherer,[2,3] is big data, that is, the data-driven decision making. The data miners who powered Barack Obama's campaign "created algorithms for predicting the likelihood that someone would respond to specific types of requests to accomplish each of those goals." These algorithms included targeting voters through complicated modeling, identifying and utilizing voter preferences to fundraising, delivering advertisements, or selecting mobilizing channels. For example, the Obama campaign (1) raised millions of dollars by making use of movie stars like Sarah Jessica Parker and George Clooney, who are the most popular among women aged 40 to 49 on the East and West Coasts, respectively; (2) bought advertising time in nontraditional TV programs to cover women below age 35 in Dade County, Florida (e.g., *Sons of Anarchy*, *The Walking Dead, Don't Trust the B – in Apartment 23*); (3) answered questions and received phone calls on the social news website Reddit in the swing states. There has recently been reported of 2016 White House bid that Hillary Clinton hired Stephanie Hannon[4] and Teddy Goff[5] to serve on her campaign team (March 2015). It is worth noting that the two people have high visibility in the big data related industrial community. Hannon, the first ever woman to hold the title of chief technology officer in presidential campaign, was the lead product manager for Google Wave. Goff, a top adviser to Hillary Clinton's campaign, played a key role in the backroom number

[2]http://swampland.time.com/2012/11/07/inside-the-secret-world-of-quants-and-data-crunchers-who-helped-obama-win.
[3]http://poy.time.com/2012/12/19/obamas-data-team.
[4]http://www.cnn.com/2015/04/08/politics/hillary-clinton-google-staffer-hire-2016-election.
[5]http://www.washingtonpost.com/blogs/post-politics/wp/2015/03/24/hillary-clinton-adds-top-digital-talent-to-campaign-in-waiting.

crunchers that helped the election of President Obama in 2008. So, let's wait and see; the data may show another "miracle" – marking the first time for a woman to serve as President of the United States of America.

2.1.3 The Underused Scientific Record

While open access of scientific publications[6] still remains controversial, Gray has turned his attention to the scientific record[7] and proposed making it "a major item" [23] of future scientific research.

The scientific record is a means of providing credible proof for research published in academic journals, including (1) the precedence of an idea, the basic data, and other proof that validates findings, hypotheses, and insights; (2) pre-publication peer reviews; (3) annotations, comments, and result re-verification, reuse, and dissemination after publication; and (4) statistical results through bibliographic citations. It also includes work associated with the establishment of common nomenclature and terminology; lectures and speeches recorded at various academic conferences; intellectual property like scientific periodicals, technical reports, and patents; and even authorial profiles or editorial reputation and transparency in reporting results.

In recent years, the relationship to scientific publications of scientific evidence has gotten more critical. In February 2011, PLoS[8] revoked a published paper on transgenic cassava (*M. esculenta*) research due to the lack of data that the research group at Melbourne University in Australia could not provide to validate their results.

There are two reasons why the scientific record is going to be a major objective of scientific research. First is the size of data, which is becoming huge, from important conferences, periodicals, references, indexes, review journals, bibliographies, classification lists, and controlled vocabulary for the different disciplines. Second is that the corresponding data is becoming ever more complicated, and added to the complex data, heterogeneous database, and numerous data files are also semi-structured text, hyperlink websites, and video clips.

Mining scientific records therefore intensifies the processing of documents and information, which can produce new outcomes, or verifications in different ways, and essentially results in data amalgamation.

2.2 ADMINISTRATIVE DATA

Since ancient times data has been assembled and promulgated as an instrument of rule. As early as the ancient Egyptian[9] and Roman times, governments already had

---

[6]Albert Einstein (1879–1955) divided scientific publications into: journal papers, book chapters, books and authorized translations.
[7]http://www.nature.com/nchembio/journal/v4/n7/full/nchembio0708-381.html.
[8]http://journals.plos.org/plosone/article?id=10.1371/journal.pone.0016256.
[9]Both the Old Testament and the New Testament mention the population census implemented in Ancient Egypt.

the capacity and intention to collect and analyze copious data on population growth and economic activities. Since the end of World War II, government agencies around the world have fortified their data with official statistics on infrastructure reconstruction to assist in government planning, business investments and economic development, academic research, media communications, and to keep the active citizen informed.

In 1993, the US federal government was pressed by Congress and voters to make budget cuts. In investigating possible wasteful spending by the government, then-Vice President Al Gore launched into action a plan called the National Performance Review and submitted two reports to President Bill Clinton. In the report *Rebuild Government with Information Technology*, it was stated that with the rise of electronic media, government should enhance its own efficiency by taking advantageous of electronic media, thus the term e-government[10] came into being. By 1996, the US government directly benefited by reducing almost US$118 billion of expenditures through closing nearly 2,000 offices and retiring nearly 240,000 federal employees.

Some other countries had similar experiences. Japan set up a three-level action plan for e-government in 1993 as well. In 1994 Canada's Minister of Industry announced a strategic framework for the development of an information superhighway. And in 1996 Britain issued its Green Book on its own systematic plan for e-government.

According to the United Nations Economic and Social Council (ECOSOC), e-government functions can be classified into four categories: (1) G2G, between governments; (2) G2B, between government and business organizations; (3) G2C, between government and citizens; and (4) G2E, between government and employees. By 2002, worldwide, there were nearly fifty thousand websites established by government departments, which is a thousand times of that in 1996; and by 2005, 94% of the UN member countries had launched an e-government project.

China is constructing its information superhighway at remarkable speed and is starting to implement a series of "golden projects" of seamless networks for selected fields. For example, the Golden Shield Project for public security, Golden Tax Project for taxation information, Gold Card Project for financial exchange, Golden Customs Project for foreign trade, and Golden Key Project for education. Furthermore all levels of governments share a basic website portal.

With the development of e-government, administrative data has increased exponentially. In the United States, for example, an estimate by McKinsey in 2011[11] shows the federal government to have owned 848 petabytes of administrative data by 2009 and ranked second (only to the 966 PB of discrete manufacturing). Currently, administrative data has two paradoxical problems – misuse and disuse – one is lack of accurate data to release; the other is to be obsessed with the accuracy of the "numbers."

---

[10]e-Government is also referred to as computerized government, digital government, online government, virtual government, etc.

[11]http://www.mckinsey.com/insights/business_technology/big_data_the_next_frontier_for_innovation.

2.2.1 Open Governmental Affairs Data

Open access as it originated from Western academics was an attempt to break the monopoly of commercial publishers in the scholarly journals market. In November 2004, the US Congress approved a bill allowing open access for the National Institute of Health (NIH), whereby each sponsored researcher would need to voluntarily submit an electronic version of his or her final peer-reviewed manuscript to PubMed Central (PMC) and allow the research to be freely available to the public. But the NIH mandate came after a long struggle. Both the open access repositories and the opposition worked tirelessly to persuade Congress to act in their behalf. Finally, in December 2005 there was passed the CURES Act (S.2104) proposed by the then Senator from Connecticut Joe Lieberman. The Act made open access compulsory and penalizes any noncompliant researcher by refused government funding.

Open access is equally targeted at public administrative data archived by the government. In December 2007, Tim O'Reilly, the founder and CEO of O'Reilly & Associates (now called O'Reilly Media), gathered together over 30 advocates, and led by Joshua Tauberer, an undergraduate student at Princeton University, they called for endowing open access administrative data with eight principles. That is, the data must be (1) complete, (2) primary, (3) timely, (4) accessible, (5) machine processable, (6) nondiscriminatory, (7) nonproprietary, and (8) license free. They also addressed more contentious issues of originality, safety, authenticity, timeliness, quality, granularity, and value of data.

This mandate led to the Obama administration's Open Government Directive (01/21/2009), required each executive department to identify and publish online in an open format with 45 days, including at least three high-value data sets, via the new federal portal at Data.gov. As of March 2015, the US government departments have provided open access to 124,200 data sets categorized into 14 topics. Data.gov aims to release data to the public and encourage innovation. For example, SimpleEnergy is a mobile application using two data sets[12] that compare a person's electricity energy consumption to his or her neighbors', and has made US\$8.9 million up to now. The action of open governmental affairs data was quickly followed by more than 30 other countries, including Britain, Canada, Australia, and Korea. To put this into perspective, in December 2011 the United States and India had announced cooperation in rebuilding Data.gov into an open source platform; by 2012, all codes were open to the public – anyone in the world can now download, use, and modify the open data platform.

In the United States at a district level, San Francisco, being one of the most popular locations for company start-ups, chose to launch open data on crime, geo information, and a 311 information hotline (a nonemergency municipal administration service), totaling about 200 open data sets. So far there have been over 100 data products developed in a number of areas using San Francisco's data, including public security, transportation, entertainment, and environmental protection. For instance, Stamen Design created a website called Crimespotting, an interactive map for visualizing criminal records with real-time updates, which situates crime scenes to alert tenants or

---

[12]One is US census data, the other is electricity power sales, revenue, and energy efficiency data.

people searching for apartments and also those who go home late at night. Currently, David Chiu, president of the San Francisco Board of Supervisors, has been pushing for a legal change to allow government to collect 1% to 3% of transaction costs from companies that use this data.

2.2.2 Public Release of Administrative Data

Besides data acquisition, another instrument of government is to regularly release data on changes in the national economy, social services that affect people's livelihoods, census results and statistical bulletins, price indexes, industrial standards, and so forth. Administrative data of a country is released through the Internet, and the target audiences not only include all of its citizens but also governments of other countries and international organizations. If the data is deficient or data quality is low, the governmental body may be subject to frequent interrogation. Government data thus not only relates directly to the formulation of public policy, it impacts national competitiveness and international relationships.

Take the review of the UN Climate Change Conference of December 2009 in Copenhagen,[13] according to which, based on the principle of common but different responsibility defined in the Kyoto Protocol, there was an accusation that the accumulative greenhouse gas emissions by China and the United States account for one-third of the global total, although China as a developing country was enjoying exemption. The Chinese government was so embarrassed that China's Prime Minister had to declare that by 2020 carbon dioxide emissions per unit GDP (gross domestic product) will be reduced by 40% to 45% compared to 2005 emissions levels. Hypothetically, China, the largest developing country, could provide evidence to show that the final manufactured products consumed by the West are the cause of the high carbon emissions, instead of just adopting a diplomatic approach. Would that approach be more positive?

Apart from disabling all the false or bogus data, maintaining government secrecy, and not offending a third-party interest, government departments have still to improve the timeliness of administrative data releases in the form of official files or reports. Fortunately, we can analyze the data of other fields to check data authenticity in a designated field, and simultaneously improve the timeliness of the results. Give a comparison of the US Centers for Disease Control and Prevention (CDC) and Google.[14] Under normal circumstances, the CDC often delays the announcement of new influenza cases for a week or two. One day, the Googlers guessed that the 50 million most frequent search entries might be targeted at getting epidemic disease information. They designed algorithms to process millions of hypotheses and test the search entries, comparing their results against actual cases from the CDC, and then preemptively published an influenza forecast paper in the *Nature* [25] a few months before the H1N1 struck in 2009. This case proved that mining the data of other fields can be helpful to a certain extent in prompting government release of information.

---

[13]The full name of Copenhagen Conference is "the 15th session of the Conference of the Parties to the UNFCCC and the 5th session of the Conference of the Parties serving as the Meeting of the Parties to the Kyoto Protocol," held May 7-18, 2009.
[14]http://www.nature.com/news/2008/081119/full/456287a.html.

2.2.3   A "Numerical" Misunderstanding in Governmental Affairs

Worldwide, almost all civil servants believe that their jobs consist of routinely crunching numbers in working up budgets, funding requests, and assessment reports, and even in preparing written briefings, memos, and meeting materials they still need to add a lot of numerical evidence. Furthermore the resulting numbers are applicable for only a limited period of time.

Regrettably, there is no respite. The new big data buzzword is "datafication," according to Mayer-Schonberger, for all the government administrative data not collected in numeric form [1, 26]. Datafication is expected to be completed in two phases: (1) datamation by human information processing and (2) digitization by the machine to convert the processed "information" into a digital format "data," expressed by binary code with 0 and 1. In other words, the real word is quantified to be information through datamation and then digitization expresses the information as data.

By this token, the numeric is a very small sample of administrative data. The same as most of the time we cannot see the forest for the trees, it is unwise to rely on tiny samples. Simultaneously, we need design a set of assessment strategies to verify the results even if each data object of the sample is correct. But, as pointed out by Polish statistician Jerzy Neyman, no matter how perfect the verification assessment, there are inherent flaws in the method of purposefully choosing samples.

At the 2011 first International Conference on Natural Language Processing (NLP), when Michele Banko and Eric Bill from Microsoft Research presented their paper "Mitigating the Paucity-of-Data Problem: Exploring the Effect of Training Corpus Size on Classifier Performance" [27], they showed how an algorithm that performs the worst on a small data set significantly improves with more data, and is almost equivalent to the best algorithm for a big data set. Anand Rajaraman's @Walmart-Labs 2012 Netflix[15] competition concluded with a similar observation by the winner who had a simple rudimentary algorithm and an additional data set of 18,000 movies.

The time is now for government departments to do away with their obsession with the accuracy of the "numbers." Algorithms exist that can produce reliable, affordable, and quick results from all available data the government already has.

2.3   INTERNET DATA

The concept of the Internet originated from Joseph Licklider's "Intergalactic Computer Network" that "started at Sun [and] was based on the fact that every computer should be hooked to every other computing device on the planet" [28]. Almost half a century later, the World Wide Web, originally designed for military and large-scale scientific research, merged physical networks of different type, different scale, and different geographic areas into one by integrating triple play,[16] mobile network, the Internet of Things, and cloud computing.

---

[15]http://allthingsd.com/20120907/more-data-beats-better-algorithms-or-does-it.
[16]https://en.wikipedia.org/wiki/Triple_play_(telecommunications).

2.3.1   Cyberspace: Data of the Sole Existence

In 1984 William Gibson's novel *Neuromancer* swept science fiction's "Triple Crown": the Nebula Award,[17] the Hugo Award,[18] and the Philip Dick Award,[19] and the author dubbed "Father" of Cyberpunk.[20] Other famous cyberpunk writers include Bruce Sterling, Rudy Rucker, Pat Cadigan, Jeff Noon, and Neal Stephenson. The world of cyberpunk is an anti-utopian world of despair, like Film Noir, mainly depicting social decadence, artificial intelligence, paranoia, and the vague boundary between reality and some kind of virtual reality. Novels of this genre have had a large impact on linguistics and popularized certain technical terms, like cyberspace, intrusion countermeasures electronics, and virtual reality; all the while they tend to be disdained and ignored by the majority of readers. Yet, many young people were fascinated by the hardware architecture of the Internet during its sprouting stages of growth.

In 1999, two movies, *The Thirteenth Floor* and *Matrix*, demonstrated existentialism of "data" as the only thing in a virtual world. So the game tycoons began thinking of how to make money from virtual worlds. In 2003, the virtual world was finally created in a true 3D online game *Second Life*, which was launched by Linden Lab, a game developer in San Francisco. *Second Life* was not good at the operability of characters, authenticity of movement, and the refinement of color, character, and scene, compared with the role-playing virtual games like *World of Warcraft*, *Perfect World*, and *Soul Ultimate Nation*; however, it was unique for two reasons:[21] first, the participants in *Second Life* – call residents – do not follow a directive from the creators of the game; second, it is the first time in this virtual world that a game would allow residents legally own the content they create (landscape, vehicle, machine, etc.), providing an incentive for entrepreneurship. What's more, the *Second Life* economy mimics the real world economy in that there is an official currency, the Linden ($L). The exchange rate of the Linden is basically stable, about 300 Linden dollars for one US dollar, and the currency may be bought and sold on *Second Life*'s official currency exchange. A small section of residents earn hundreds of or even thousands of US dollars per month. For example,[21] the man named "Elliot Eldrich," a semiretired engineer lives his real life in Spokane, Washington. He adds to his savings in this virtual environment as a developer, earning about US$500 per month in the process. By 2007, the number of the "residents" in *Second Life* approached 40 thousand. CNN set up a branch, IBM started a marketing center, countries like Sweden established embassies, and more than 300 universities including Harvard use it as a teaching method.

---

[17] Nebula Award was founded by Science Fiction and Fantasy Writers of America.

[18] Hugo Award was founded by World Science Fiction Society in memory of Hugo Gernsback, who is known by some as the "Father of Science Fiction". It's officially called "The Science Fiction Achievement Award" and is considered comparable to the Nobel Prize in the science fiction circle.

[19] Philip Dick Award was founded by Philadelphia Science Fiction Society in memory of Philip Dick, a science fiction writer.

[20] http://en.wikipedia.org/wiki/Cyberpunk.

[21] https://www.urbaninsight.com/articles/2ndlife.

Just as the creator of *Second Life* Philip Rosedale said, "I'm not building a game. I'm building a new continent." We want to say, no matter how far away this "continent" is, we can arrive in a moment. And in this virtual realm – data – the only existence, is the largest resource and asset.

2.3.2 Crawled Fortune

In October 1994, Sir Tim Berners-Lee, founder of the World Wide Web, established the World Wide Web Consortium (W3C) at MIT and released nearly a hundred relevant standards. Since then, the web has become the largest source of public data in the world and has brought together various types of data involving news, commercials, finance, education, governmental affairs, and business.

Almost at the same time, the first network crawler was born at MIT.[22] Web crawler is a kind of "small insect" that can "grab" data from websites at the speed entirely out of human reach with an epoch-making significance, due to the effect of search engines that evolved from it. Each crawler sends a request and indicates its identification (a field user-agent is used to indicate the identification in the request) when it crawls the web for hyperlinks, page content, or logs. For instance, the identification of Google's web crawler is GoogleBot; Baidu's, BaiDuSpider; and Yahoo's, Inktomi Slurp. Website administrators know from the log about which and when a web crawler visited, how many pages to be copied, and so forth.

In order to analyze the huge quantity of "crawled" data, web mining [29], a kind of mining technology that is different from information retrieval and extraction, has emerged and mines the content, structure, and usage of the web to search information and obtain implicit knowledge used for decision making.

However, some relevant third-parties have been taking these "crawled" data as their own fortune. According to Google's official report in 2008, it had obtained over 1 trillion web documents from the Internet through this method. Albert-László Barabási, founder of the global complex network research authority "scale free network," said in his new book *Bursts: The Hidden Pattern behind Everything We Do, from Your E-mail to Bloody Crusades* [30] that Google's famous "Don't be evil" philosophy "covers a huge intellectual black hole." The firm "uses its billions of dollars to sweep up the best engineers and scientists in their fields, who are then locked up in its Santa Clara Googleplex, where they are shielded by strict nondisclosure agreements so that they can rarely publish their findings."

2.3.3 Forum Opinion Mining

Internet forum, also called message board, mainly provides storage spaces for online discussion; technically, it is a replacement of the earlier bulletin board system (BBS) service. There are various topics discussed via forums, such as life and society, news, education, tourism, and recreation. Some forums are all-inclusive, while others focus on specific topics. In 2010 iResearch carried out a survey in China of discussion

---

[22]Web crawler is also called "web spider", and was designed by MIT student Matthew Gray in June 1993.

topics:[23] life topics, and interactive recreation topics accounted for 19.9% and 19.4%, respectively.

Forum data is typically text-heavy and unstructured, but users also share music, videos, and photos, wherefore sometimes contains multimedia. Forums are often oversized by moderators, who monitor the interchange of contributors and make decisions on content and the direction of threads. Common privileges of moderators include:[24] deleting, merging, moving, locking, and splitting of posts and threads; renaming, banning, suspending, and warning the members. Today, a big challenge for moderators is the problem of keygens, advertisements, and spams. Many forums deal with this by blocking posting based on keywords or deleting posts through manual browsing.

Opinion mining, also known as sentiment analysis,[25] is to identify and extract subjects or opinions on text descriptions in materials; related tasks include [31] opinion search (search for a specific subject or opinion with object features), opinion classification (judgment of semantic preferences), feature-based opinion mining and abstracting (feature extraction), comparative sentence mining and comparison, and opinion fraud detection (similarity and dissimilarity between contents, detecting grade, and content exception). Thus using opinion mining technology may partially relieve the workload of data maintenance for forum moderators.

2.3.4 Chat with Hidden Identities

In our real life society, identities like costumes, tie us to others "like us" and may offer us stability, direction, and "a sense of belonging within that society." However, this superiority that comes from identities, in turn, may prevent our unique individuality from fully and clearly expressing because of the limitations of the identities themselves.

Instant messaging, a kind of online chat, offers mostly a real-time text transmission with geographical position (if a location-based service is used) between senders and receivers on the Internet. Earlier applications of instant massaging include ICQ, AOL, MSN Messenger, Skype, and QQ. Currently, with the development of fixed-mobile interconnection, new applications like Whatsapp, WeChat, and Line emerged. Recently, WeChat claims 5 billion active users, with 1 billion outside China.

The convenience of using instant messaging is to expand the social networks of both senders and receivers, with an identity barrier, which is to say, we are not our identities. As user identity is hidden, each "virtual" person's opinion, emotion, and social relationship become more real during the communication, even quite a few scholars believe that the chat data set is an alternative to traditional telephone to study interpersonal interactions of the socioeconomic reality. This is because, as both the size and type of active data increase, despite the hidden identity, correlation analysis

---

[23]Quoted from: 2010 Research Report on Chinese Online Communities. http://news.iresearch.cn/Zt/128854.shtml.
[24]http://en.wikipedia.org/wiki/Internet_forum.
[25]http://en.wikipedia.org/wiki/Sentiment_analysis.

based on cross-checking of data can mine the anonymous information. Take QQ Circle, for example. Tencent integrates various relationships in chat data, such as second-degree friendships, group memberships, classmates, and blog audiences, and it is that easy to outline a user's circle of friends.

2.3.5   Email: The First Type of Electronic Evidence

To realize the sending and receiving of files through the Internet, Ray Tomlinson invented electronic mail (email) in 1971. As a vital application of the Internet, email has a lot of merits, such as low cost, rapid delivery, easy storage, and barrier-free passage worldwide. Thus email has effortlessly taken the place of traditional correspondence. The email address is composed of three parts: the domain name of the server of user's mailbox to send and receive mail, the account number of user's mailbox (i.e., each user gets a unique identification from the same mail-receiving server, and the separator symbol "@").

When most major email providers, including Yahoo! Mail, and Hotmail, started charging a setup or monthly for email, Google introduced a free web mail service – Gmail on April Fool's Day in 2004. Gmail is different enough to be the redefined email for the following reasons. First, the Gmail invitation system forever changed the startup way (i.e., Google first gave out 1,000 accounts to Google employees, friends, and family members to try the new email system), and new users had guided access via an invitation from existing users. Second, Gmail brought three entirely new user experiences: (1) no need to delete email, as the free storage space increases from an initial 1 GB to over 10 GB; (2) searching without sorting, whenever email is sent or received, it can be found with search engine; (3) email automatic grouping, in other words, all relevant emails will be carried in the corresponding context because of their relevance. Despite the famous AJAX[26] application offers a convenience praised by users, Gmail has still been criticized by controversial issues such as automated scanning and incomplete removal of mails (i.e., Gmail automatically scans emails to add context-sensitive advertisements and to filter spam, and only marks the email in background servers instead of physically removal, even if user deleted a mail).

Email data is composed of multiple types of data, including text, graphics, motion picture, audio, and video. In the February 2009 issue of *Science*, fifteen researchers [32] from several the US universities and institutions argued that analyzing email data has been an "emerging research field" in that the mail may reveal patterns of individual and group behaviors, such as "patterns of proximity and communication within an organization" and "flow patterns associated with high individual and group performance."

Email data is the first undeniable electronic evidence. In 1993 the UNCITRAL (the United Nations Commission on International Trade Law) finalized the *Draft Uniform Rules on the Legal Aspects of Electronic Data Interchange (EDI) and Related Means of Trade Data Communication*, which had recognized that contracting based on email

---

[26]AJAX refers to Asynchronous JavaScript and XML. It's not an abbreviation but rather is a new word created by Adaptive Path cofounder Jesse James Gaiiett.

exchange is identical to "sign an original copy in written form." As an electronic evidence, the email should be handled by the following issues: (1) the authenticity of sender and receiver, which are supported by the Internet service providers; (2) the authenticity of email content including header, body, and attachments, which are supported by metadata as timestamp; (3) the primitiveness of email, which involves three steps, the first step being to (a) check the differences between the evidence provided by both the claimant and respondent, then (b) collect the evidences using a CD-ROM (compact disc read-only memory) disk to guarantee no modification of the content, and after that, (c) check the CD-ROM disk on the spot, and finally seal the disk with a signature by all parties including a court operations officer.

2.3.6   Evolution of the Blog

The blog (short for weblog) evolved from Internet forum to create an online diary, in which people would keep a running account of their personal lives. Hosted by dedicated blog hosting services, the blog rapidly become popular and spread from the late 1990s to now. According to Technorati (a real-time search engine dedicated to the blogosphere) Chairman David Sifry's blog,[27] there are "14.2 million weblogs in July 2005" and "as of July 2005 over 80,000 were created daily – a new weblog is created about every second." However, the blog is not static; it is, rather, evolving (as shown in Figure 2.1).

The left clade in the figure represents the variation in the blog's content. The blog in the initial state was mostly unstructured text and contained few images, hyperlinks of other blogs or websites, and media items. Then it evolutes to focus on

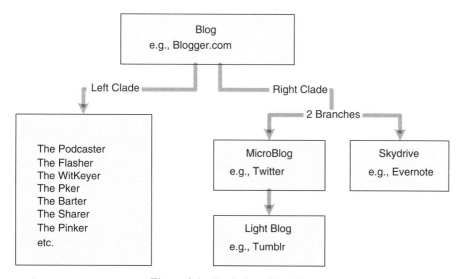

Figure 2.1   Evolution of the blog

---

[27]http://www.sifry.com/alerts/archives/000332.html.

professional fields such as videology and photography, and certain other thematic items such as PK and bartering. Simultaneously, a group of typical bloggers emerges, among them the podcaster (people keen on video playing), the flasher (people keen on making Flash), the witkeyer (people keen on converting knowledge or experience into revenue), the sharer (people keen on exposing their private lives); there are even the pinker, the xuankr, and the huankey stocked in the "Chinese version."

The right clade has two branches: in the first branch, the text-heavy blog cuts its length to "allow users to exchange small elements of content such as short sentences, individual images, or video links,"[28] including the Microblog (e.g., Twitter) and the Light blog (e.g., Tumblr); in the second branch, the blogs retain the store function and evolute into the Skydrive, such as Evernote.

2.3.7   Six Degrees Social Network

John Barnes from the University of Cambridge coined the term "social network" [33] in 1954. Much later in 1967, Stanley Milgram of Harvard established the theory of six degrees of separation, which assumes that any two people not knowing each other anywhere in the world can be made to connect together in a maximum of six steps (one step, as "one degree"). Then in 1998, Duncan Watts with his mentor, the famous mathematician Steven Strogatz, published a paper called "Collective Dynamics of 'Small-World' Networks" in *Nature*, which proved the six degrees of separation, and established the small world theory. Nicholas Christakis, a Harvard professor who shifted his field of research from medicine to sociology, found in 2009 [34] that connections within three degrees contain strong ties that can induce action, whereas connections beyond three degrees are weak ties that can only transfer information. Here, a tie [29] is said to exist between communicators in which they exchange or share resources like social support or information.

Today, various social networks have attracted billions of people to post information about their private lives. The large-scale new data containing characteristics associated with human groups not only is a gold mine for data scientists, it also attracts experts from other fields such as psychology, sociology, and journalism.

While the billion-user Facebook relies on ads to earn money, LinkedIn has begun "gold rushing" from the social network data. The business model is to provide professional positions and social network for the individual user: (1) strong ties within three degrees are free, meaning users may view the networking within three degrees, such as establish contact and start business communication; (2) weak ties beyond three degrees are charged with a fee, meaning users need to pay certain amounts if they want to view a network beyond three degrees. Although these premium subscriptions had generated 20% of total revenue at year end in 2012,[30] it was more than enough to get the attention of Wall Street investors.

[28]http://en.wikipedia.org/wiki/Microblogging.
[29]http://emmtii.wikispaces.asu.edu/file/view/Shakespeare%20Hero%20Demo.pdf.
[30]http://heidicohen.com/linkedin-where-the-money-is-and-11-facts-to-prove-it.

2.4   FINANCIAL DATA

Wide differences exist in the definition, scope of research, and integration of a strong or weak correlation in the field of finance. But they all share a common understanding that significant research on finance is possible only when the collected data is complete and accurate. A bank deals essentially not about the money but information, noted William Rhodes, the former Senior Vice-Chairman of Citigroup and Citibank. The data is the only crucial thing in finance, and it is the mainstay in managing currency derivatives, closing financial agreements or contracts, processing banking businesses, and especially in controlling the strict timeliness of high-frequency and ultrahigh-frequency transactions.

2.4.1   Twins on News and Financial Data

The Burton-Taylor 2014 financial market report stated,[31] in 2013, that "global spending on market data analysis showed the weakest growth since 2009, up 1.10% to US$25.88 billion," and "Bloomberg market share reached 31.71%, while Thomson Reuters commanded 26.93% [reflecting divestiture of their Corporate/IR and selected other businesses]." Inarguably, the America's Bloomberg and Britain's Thomas Reuters are the top two news agencies and also the world's leading financial data providers.

Some people, accordingly, noticed and started to prove the potential correlation between news and financial data. For example, the news items of CNN "Libyan forces sock oil port as global pressure' (03/11/2011)" caused a fall in stock price for oilfield service enterprises; similarly, in the wake of the Japan's Fukushima Daiichi nuclear disaster,[32] stock prices of several energy enterprises reliant on nuclear sources and listed on stock exchanges plummeted while renewable energy companies rose sharply in value. For these reasons, money managers use association analyses to extract event connections in order to forecast trends in financial markets.

2.4.2   The Annoyed Data Center

The ever-widening discrepancy between CPU performance and that of storage subsystems has spawned the huge growth in demand for data centers. James Porter, the first full time digital storage analyst, said that "the capacity of digital data storage worldwide has doubled every nine months for at least a decade, at twice the rate predicted by Moore's law for the growth of computing power during the same period."[33] In the United States, the federal government opened 2,094 data centers in 2010, an increase of 3.85 times compared with the 432 centers in 1998. Furthermore one of the lessons from the "9/11" attacks was the need for financial services companies, especially the banking industry, to have a backup data center "located away from

---

[31] http://www.burton-taylor.com/samples/B-T_Global_Market_Data-Analysis_5_Year_Competitor_
and_Segment_Product_User_Institution_Analysis_2014-Information_Kit.pdf.
[32] http://en.wikipedia.org/wiki/International_reactions_to_the_Fukushima_Daiichi_nuclear_disaster.
[33] http://sceweb.uhcl.edu/boetticher/ML_DataMining/p28-fayyad.pdf.

their primary data center." Thus, after 9/11, data centers were rapidly constructed. In China, according to Shanghai Securities News, by the end of the 2011 financial year, 255 banking organizations of China had constructed 501 new data centers, with a total investment amounting to 69.87 RMB billion, a year-on-year increase of 38.3%.

However, building a data center is not straightforward. Apart from incredible construction costs, data centers have very strict requirements as to environment, energy supply, and other specifications. For example, in the site selection for the data center, there are multiple factors to consider: (1) the distance factor, essentially a straight-line distance from the firm's headquarter; (2) the climate factor, whether the weather is cold, dry, and far away from natural disaster zones such as earthquakes and volcanoes; and (3) the communication factor, namely how to connect conveniently the sealed or land cabling. Simultaneously, factors of land and resources cannot be ignored in constructing a data center. The increasingly intense contradiction between the supply of and demand for land and resources has urged local governments to consider land price, productive value, revenue, and environmental protection. In addition, a data center is power-hungry in use, according to Energy Star Program's 2007 report,[34] "in 2006, US data centers consumed an estimated 61 billion kilowatt-hours (kWh) of energy," and "if current trends continue, by 2011, data centers will consume 100 billion kWh of energy, at a total annual cost of US$7.4 billion."

2.5 HEALTH DATA

Benefited by the growth and development of health and biomedical informatics, health data has unmatched value in a quickly evolving complex healthcare marketplace. This is because a retrospective look at what actually happened is conducive to improving the collective experience of human healthcare services.

2.5.1 Clinical Data: EMRs, EHRs, and PHRs

The United States is an internationally recognized leader in the research development of health information systems. These systems include the 1960s hospital management information system (HMIS), the 1970s clinical information systems (CIS), and the 1980s picture archiving communication system (PACS), and the best today are the HELP system at Intermountain LDS Hospital, the COSTAR system at Massachusetts General Hospital, and DHCP system of the Department of Veterans Affairs.

In 1991, the Institute of Medicine (IOM) began researching a computer-based patient records (CPRs) system. The IOM published a book *An Essential Technology for Health Care*, and joined with the Markle Foundation and 13 health and IT organizations to put out a proposal to launch an electronic medical records (EMRs) system guided by a government uniform standard. The idea was to ensure that timely information on patients could be transmitted electronically to reduce medical error.

---

[34]http://www.energystar.gov/ia/partners/prod_development/downloads/EPA_Datacenter_Report_ Congress_Final1.pdf.

In July 2003, the Department of Health and Human Services adopted two measures to promote a national EMRs system. One was to purchase the license of a systematized nomenclature of medicine (SNOMED) so that it could be used all over the United States without an additional fee. The other was to designate a health institute to design and develop a standard EMR template so that it could be provided to other institutions free of charge. At the end of 2003, then-President George W. Bush signed the Medicare Prescription Drug Improvement and Modernization Act that required the Centers for Medicare and Medicaid Services to prepare an e-prescribing standard as a first step. In 2009, in order to mitigate the Sudden Recession, the United States reserved US$36 billion from the American Recovery and Reinvestment Act funds for the EMRs program.

Until recently, worldwide heath institutions had extended the scope of digitization to electronic health records (EHRs) by using standardized terminologies and ontologies as data resources. Currently, data analysis in this field is generally text analytics according to evidence-based medicine. For example, in August 2011, Francisco Roque from Technical University of Denmark published a paper on PLoS computational biology [35] that applied text analytics on a data set collected from 5,543 patients during 1998 to 2008 at Sankt Hans Hospital, and classified patients with symptoms related to diseases. But a persistent problem in this field is that health institutions sequester patient data behind a data isolation wall. According to a report [35] released by PwC Health Research Institute in 2011, only 14% of patients has access to his or her own personal health records (PHRs).

2.5.2 Medicare Claims Data Fraud and Abuse Detection

Owing to the diverse diseases, treatments, and services, there are many ways that Medicare fraud or abuse can occur, all with the same objective:[36] to illegally extract money from the Medicare program. In the United States, as announced by the Department of Justice, discovery of the largest medical insurance fraud case in US history took place in July 2010, when 94 suspects were apprehended, including doctors, nurses, clinic owners, and managerial staffs, for appropriating US$251 million by fraudulent claims. But some records are meant to be broken. In February 2013, the Federal Bureau of Investigation carried out a surprise attack on a Texas-based SCOOTER store,[37] which was accused of providing medically unnecessary scooters and suspected to have received US$723 million in medical insurance claims. In France, an anti-fraud executive board was established in 2008 to investigate medical insurance fraud and abuse cases. From 2008 to 2010, as this initiative gained strength, the number of investigated claims exceeded 5 million euros, half of which were proved to be illegal.

Besides revealing health facts about individuals and the focus population, claims data can show whether Medicare fraud or abuse exists. But, before

---

[35]http://e-patients.net/u/2011/02/PH-11-0101-Putting-the-patients-into-meaningful-use-PwC.pdf.

[36]http://en.wikipedia.org/wiki/Medicare_fraud.

[37]http://www.cbsnews.com/news/the-scooter-store-shutting-down-after-federal-scrutiny-cbs-probe.

ever more vast amounts of claims data are accumulated, identifying Medicare fraud or abuse mostly depends on designing good measurable points to use as fraud or abuse indicators. The mining techniques that we currently have need to include deviation detection, idiosyncratic clusters (like abnormal groups or peculiarity groups) [36], and categorically odd claims [37], besides the usual excessive medical prescriptions, high rates of clinical tests, and frequent Medicare card use.

Claims data combined with clinical data may provide a better answer. In September 2012, a startup company called Predilytics applied machine learning to the field of medical insurance and raised US$6 million from Flybridge Capital Partners, Highland Capital Partners, and Google Ventures. Predilytics claims to have an algorithm capable of using available data (including medical insurance claims, medical prescription, clinical test, conformity certificate, call center, EMR, and nursing procedure data) that, without manual intervention, can detect deviations much faster and with more precision than a traditional rule-based statistical regression.

2.6   TRANSPORTATION DATA

The automotive industry developed rapidly. The early automobile featured three wire wheels (which differed from the wooden wheels of carriages), and as manufactured by Karl Benz, was only capable of 18 km per hour. In comparison, in our time sports cars can accelerate from zero to a hundred in 2.5 seconds. Global vehicle ownership has jumped from 250 million in 1970 to over 1 billion in 2010, and is expected to exceed 2.5 billion by 2050. However, an attendant problem is traffic congestion. Despite continuing to improve road conditions and traffic rules, the biggest challenge of each city is still how to cut traffic jams.

Technically, a transportation system can be regarded as a supply–demand relationship composed of three factors: consumer, vehicle, and roadway. To mitigate traffic congestion, cities usually have to (1) intervene by raising parking fees in the central city zone, by levying a traffic congestion charge, by asking employers to implement flexi-time and staggered work hours, by restricting certain license plates or cordoning traffic access to the urban core, by encouraging use of public transportation and carpooling; (2) enlarge traffic areas, add shopping malls, provide more comprehensive or expand public transportation facilities, as well as the transport infrastructure such as bus bays, U-turn highway lanes, ring road entries, broad ramps, fast roads, truck highways, bridge roads, overpasses, subways, and light rails.

Intelligent transportation systems are being advanced to provide integrated traffic service management capable of real-time, accurate, and highly efficient functioning. The systems that implement data acquisition are only part of this development; the aim is to enhance the mining of transportation data. Transportation data can be roughly classified into three types: trajectory data collected from on-board equipment, fixed-position data collected from roadside equipment, and location-based data collected by geographic and satellite technologies.

2.6.1 Trajectory Data

Trajectory data records a motion path of floating objects such as a vehicle, a vessel, or an aircraft, directly or indirectly. It also reflects the subjective will of the driver and environmental restrictions during a floating object travel. Trajectory data initially depended on on-board equipment with a GPS tracking device for collection. Such on-board equipment is classified into private type (only having a navigation function), general type (with multiple functions of navigation, security, and entertainment), pre-assembled type (e.g., OnStar, G-book, ATX) or post-assembled type (e.g., AutoNavi). With the development of telematics, technologies of these equipments include, but are not limited to, WLAN (wireless local area network), CC (cellular commemoration), DSRC (dedicated short range communication), GSM (global system of mobile communication), and DMB (digital multimedia broadcast).

Currently, all buses and taxis are equipped with on-board electronic equipment, since both types of vehicles travel all day and contribute to much of overall urban traffic. Compared to buses that frequently stop at stations, taxis are more advantageous positioned in acquiring a stable data set. This is because taxis travel in all weather conditions, drive a uniform type of vehicle similar to that of other vehicles on the road. An analysis of taxi trajectory data does not require conversion due to vehicle model and crowd differences.

2.6.2 Fixed-Position Data

Fixed-position data refers to transportation data collected by an embedded type or a nonembedded type of detector installed at a roadside. The roadside equipment may be (1) a ring coil detector, an embedded detector buried under the pavement of the road that registers vehicle speed, traffic, and time occupancy through changes in the coil's magnetic field; (2) a vapor pressure type detector, a hollow rubber tube laid on the pavement of road that detects quantity of vehicles and speed through air pressure changes when vehicles pass by; (3) an infrared detector installed on a bracket, over a bridge, on a bridge girder, or on the heel post of a bridge, detects vehicle speed using a contrast of the infrared radiation energy by the pavement and vehicle; (4) a microwave or radio wave reflective detector installed on an upright column higher than road shoulder to detect vehicle speed, traffic, and vehicle intervals by emitting a fixed frequency or frequency-modulated continuous wave; (5) a sound wave (or supersonic-wave) detector that is normally small and placed on top of the driveway to register traffic, vehicle speed, or driveway occupancy of driveways by comparing different voice signals and distinguishing among different types of vehicles; (6) an automatic video detection device installed on top of a driveway or on the road side to feed into processors analyzing the changing characteristics of the video image (e.g., background gray value), such as vehicles pass; (7) a Bluetooth transportation detector that "calculates travel time and provides data for origin and destination matrices using the MAC (media access control) addresses from Bluetooth devices,"[38] like mobile phones, headsets, and navigation systems in passing vehicles.

---

[38] http://self.gutenberg.org/articles/intelligent_traffic_system.

In effect, due to the fact that the numerous types of equipment are continuous real-time monitoring and failure-prone, fixed-position transportation data is mostly a data stream featured by high dimensionality, heterogeneous data, and online quickly reach.

2.6.3   Location-Based Data

Location-based data represents series factors of a natural or social phenomenon with geographic positions and distribution characteristics. The data includes, but is not limited to, remote sensing data, geographic data, atmosphere and climate data, land cover type data, geomorphologic data, soil data, hydrologic data, vegetation data, residential data, river data, and administrative boundary data. Here, we discuss the two most important ways in which geographic data is obtained: (1) by remote sensing whereby the landmark target data is collected from the upper air using long-range sensitive, reflected, or self-emitted electromagnetic waves, visible light, or infrared light, and (2) by direct or indirect association with various geographic phenomena.

The way location-based data is stored is to convert it into geographic coordinates for the desired geographic coordinate storage system: (1) as spherical coordinates, also called latitude and longitude, for reuse as a geographic coordinate, in the form of latitude and longitude, to define a spatial location on a three-dimensional spherical approximation of Earth; (2) as projection coordinates, also called plane coordinates, for reuse on a three-dimensional irregular ellipsoid approximation of Earth, but only after the spherical coordinates were projected and converted into a plane for storage. Mercator projection, a common projection coordinate, is a cylindrical equal-area projection created in 1569 by Gerardus Mercator, a Flemish geographer and cartographer. The Mercator projection is on a hollow circular cylindrical approximation of Earth (and tangent to the equator of Earth), with a lamp at Earth's inner core to project spherical graphs onto the cylinder. So, when expanded, the cylinder gives a world map with a value for the latitude line of zero degrees. Until recently almost all online maps adopted the Mercator projection technique, including Google Map, Bingmaps, Mapabc, Mapbar, ArcGIS online, and Baidu Map.

In effect, location-based data attempts to represent a spatial location in three ways: (1) by position, a location point on Earth abstractly converted into topological spatial relations of point, line, and plane in geometry; (2) by characteristic feature, the properties of land quality, soil texture, desertization, soil erosion, topography, pollution degree, population density, and traffic flow; (3) by time, to show the process of change with time on the spatial location. As a result location-based data mining issues involve high-dimensional images, hyperspectral imaging, and time series, to generate contents that can accommodate a wide range of complicated research.

2.7   TRANSACTION DATA

Transaction data is produced from the large quantity of transactions in retail and commercial matters. The original targets of knowledge discovery in database (KDD) were the commercial values contained in transaction data, and an early mining application came from commercial circulation to answer primarily retailer questions [6], such as Who are the most profitable customers? How to arrange the warehouse with limited space and funds? What products can be up-sold? What is the right coping strategy to handle "mutual exclusion of similar commodities in a market basket"? How to forecast revenue and growth over the next year? Today, transaction data additionally supports a wide range of business intelligence applications, such as customer analysis, directed marketing, shelf space arrangement, workflow management, materials distribution services, and fraud detection.

2.7.1 Receipts Data

Cash registers have pushed information-based reform of traditional retail businesses, and today digitize sales receipts. The early mechanical cash register, made by the Ritty brothers[39] in Boston in 1879, was entirely mechanical – nothing but simple adding machines, without receipts. The electronic cash registers (ECR) made in Japan during the 1960s and 1970s, as well as the PC-based POS (point of sale) released by IBM, provided a better and more convenient way to collect money and maintain real-time merchandise control, with accurate accounting and sales summaries.

Today, we cannot simply advance a judgment for a company based on whether the company sells online or offline. Converting paper receipts to digital data, and with effective utilization of high-tech, can keep a traditional retailer on the leading edge. Take Walmart, for example. This global retail tycoon,[40] operating over 11,000 stores under 71 banners in 27 countries, established a private commercial satellite system in the 1980s to maintain enterprise data exchange and guarantee accurate and timely transmission of data within headquarters, suppliers, delivery centers, and stores, in order to get a list of the best-selling goods. In April 2011 Walmart established its own data laboratory @WalmartLabs in San Bruno, 1,849 miles away from its Bentonville headquarters in Arkansas. Walmart has recently completed a series of acquisitions for accelerating data innovations by multiple SMEs (small- and medium-sized enterprises) in a total deal valued at US$300 million. Among these acquisitions are Kosmix, OneRiot, Small Society, Social Calenda, Set Direction, and Grabble. Walmart also is a good example of a company that realizes the purpose of inventory optimization based on understanding customer behavior and preferences: whether different weather conditions can cause changes in customer consumption habits, which store has the largest retail traffic on a specific holiday, and so on. @WalmartLabs has even developed a special tool called Muppet through Hadoop MapReduce to address merchandise information with locations. This tool processes the social media data (e.g., Facebook, Twitter, Foursquare) in real time, while at the same time performing several mining algorithms that combine Walmart's receipts data.

2.7.2 e-Commerce Data

In recent years, as electronic commerce has become an important tool for SMEs worldwide, commercial distributions have widened and changed from e-commerce

---

[39] James Ritty and John Ritty.
[40] http://corporate.walmart.com/our-story.

to e-business into today's e-economy. Among emerging economies like China, online retail sales rose to US\$ 589.61 billion in 2015.[41] According to a report by McKinsey Global institute [42] in March 2013, the C2C (customer to customer) market share was 90.4% for Taobao.com, 9% for Tencent Paipai, and 0.4% for eBay.com; the B2C (business to customer) market share was 51.4% for Tmall.com, 17.3% for Jd.com, 3.4% for Amazon, and 3.3% for Suning.com.

Despite many advantages, such as increasing transaction velocity, saving transaction costs, and liberated economic activities from the bondage of time and place, some people still question its technological content or innovation drive, and criticize the hoopla created by online retailers that boosts overconsumption. Take the online travel market, for example. Very few companies offer a platform that operates as a "channel provider" and disseminates the information of large number of travelers in order to win over vast customer groups by negotiating the lowest discount and the highest commission with thousands of hotels or airlines.

Unlike receipts data, e-commerce data covers not only market basket data but also "other" data through the entire transaction (i.e., before, during, and after transaction), such as releasing merchandise information, executing online marketing, providing after-sales services, and supporting the smooth development for e-commerce activities like settlement, logistics, and delivery system. Currently, e-commerce data analysis has even a higher concern that concentrates on consumer behavior, online shopping, and product optimization, under B2B (business to business), B2C, or C2C trading patterns. The methods and instruments for these are relatively unsophisticated; they are simple indicators extracted from businesses during online marketing. The indicators approximately include traffic indicators, conversion indicators, promotion indicators, service indicators, and user indicators. More specifically, the combined traffic and user indicators track the number of visitors or returning visitors, page views per capita, depths of visiting, number of entries, duration of stay, bounce rates, and repurchase rates, such that can be used to analyze buyers' behaviors, measure existing user activities, and estimate the costs of acquiring new users. What is more, there are promotion indicators like quantity of display advertisements, number of clicks or click-through rate, and return on investments that can analyze the market efficiency of sellers.

[41]https://www.internetretailer.com/2016/01/27/chinas-online-retail-sales-grow-third-589-billion-2015.
[42]http://www.mckinsey.com/insights/asia-pacific/china_e-tailing.

3

DATA INDUSTRY CHAIN

The industrial chain is a relatively independent subfield in industrial organization (IO). Inherent to the chain are two attributes: a structural attribute (partly discussed in Chapter 1) and a value attribute of an enterprise group structure. The value attribute of an "industrial chain" relates to the concept of a "value chain" first mentioned by Michael Porter of Harvard [38], in 1985. The industrial chain is considered to have four dimensions: a value chain, an enterprise chain, a supply chain, and a space chain.

This chapter gives a concise introduction to the data industry chain to help the reader better understand what the data industry is.

3.1 INDUSTRIAL CHAIN DEFINITION

3.1.1 The Meaning and Characteristics

First, the industrial chain is a conception of industrial organization in that it consists of associated enterprises spaced out along a so-called chain of value-added activities. Second, the industrial chain's collection of enterprises is intended to meet a specific demand, produce a specific product, or provide a specific service. Third, the industrial chain is connected by dynamic supply-demand activity of the enterprises forming its links that is entirely in strict chronological order.

From these three descriptions, we have a clear picture, likewise, of the data industry chain; that is, it contains enterprises that operate at different industrial levels, processing resources of different weights and complexities, and in a timely fashion, to satisfy each level of demand. The production of the data product is thus by an

The Data Industry: The Business and Economics of Information and Big Data, First Edition. Chunlei Tang.
© 2016 John Wiley & Sons, Inc. Published 2016 by John Wiley & Sons, Inc.

alliance of associated enterprises, each supplying added value. The data industry chain's supply and demand relationships are between the different cooperative enterprises developing the data resources. More specifically, in this chain there are (1) the upstream enterprises, extending toward the extremity of the information industry, that include data acquisition, data storage, and data management; (2) the midstream enterprises that include data processing, data mining, data analysis, and data presentation; and (3) the downstream enterprises that is data product marketing including pricing, valuation, and trading.

Thus the data industry chain has the following characteristics:

1. *Resource Orientation.* Upstream industrial links control data resources to determine the profit distribution of midstream and downstream industrial links. It is assumed that a leading enterprise W of the upstream industrial links owns a certain data resource S. Without W's licensing and transferring, enterprises of midstream and downstream industrial links cannot exploit and be utilized a part of S; and such dominance by W in this space can hardly be broken.

2. *Non-obvious Ecological Effects.* The non-obvious ecological effects are both internal and external. The data industry chain is integrated internally to safeguard their profits, meaning the upstream adopts a resource integration strategy for shortening total length of the data industry chain via control of the core data resources. But conversely, in having initially no data resources, the midstream and downstream must grab data resources using their own technologies or market advantages, and then extrude the data industry chain. Otherwise, they may choose to connect with other traditional industries using external data resources, and hence increase profits of themselves and another (i.e., the specific external industries).

3. *Independent Entities.* The economic entities that have formed an alliance along the data industry chain are mutually independent. For example, a stock exchange (SE) regularly provides open access to their financial data for enterprises that process data. All such entities as the stock exchange and other enterprises in this strategic alliance operate independently.

4. *Customization.* Unlike general merchandise that extracts generality from individuality for standardized processing, the data product has a "customization" feature. But the demand for customized data products originates from a rough idea due to a customer's vision without much detail provided.

5. *Intangible Products.* The data product is intangible; it is usually formed through software or algorithms and delivered through modern communications, or sometimes is just a deduction. The value of the data product is demonstrated in the application.

6. *External Dependencies.* The data industry chain is an organic whole, each part of it is independent (that guarantees the loose coupling of it) but fits well with the other parts; that is to say, the ideal situation for the data industry chain is without chasms in the upstream industrial links, without slip-ups in midstream industrial links, or without deficiencies in downstream industrial links. Yet such an ideal situation need support from the external environment,

including the establishment and enforcement (e.g., data security, data privacy, data resource protection) of related policies (e.g., enterprise-supporting policies) and legal recourses, human resources management, venture capital investments, and industrial infrastructures.

3.1.2 Attribute-Based Categories

An "attribute" provides a category means based on the concept of an industrial chain that includes resource properties, driving forces, supply-demand relationships, and market discrepancies. The structural and value attributes provide a useful way of describing the data industry chain.

1. *Resource-Oriented Industrial Chain.* Supply-demand relationships among enterprises of the internal data industry chain may be resource-oriented, product-oriented, market-oriented, or demand-oriented. Enterprises that have a resource advantage in the data industry are the dominant and influential ones. Thus the data industry chain is a resource-oriented industry chain.

2. *Dynamic Efficiency Supplied by Technology and Business.* Dynamic efficiency, a term used in economics, refers to an economy that focuses on knowledge, technologies, or businesses to build up productive efficiency over a specific time period. Despite the fact that data technology is a dominant factor in the structure of the data industry, innovating business models or adjusting operations according to business needs and preferences is important. This is why we say that the dynamic efficiency of the data industry chain is supplied by technology and business.

3. *Nondependent Industrial Chain.* According to the degree of interdependence of the upstream, midstream, and downstream industries of the industrial chain, there are three attributes: monopolistic model, competitive model, and dependent model. The monopolistic industrial link is able to control the entire industrial chain. The competitive upstream industrial links can directly handle the market and not need to negotiate interdependence issues with downstream operators. The dependent upstream and downstream industrial links may mutually be providers and customers with a high degree of interdependence. Obviously, the first two models are applicable but the third is not. So the data industry chain is a nondependent industrial chain with both monopolistic and competitive tendencies.

3.2 INDUSTRIAL CHAIN STRUCTURE

Generally, among the economic entities of the industrial chain there are resource suppliers, multilevel product providers, and terminal customers. Among the environmental elements there are industrial technology research institutes, civilian industrial bases, industrial investment funds (or other venture capital funds), government departments, intermediaries, public service platforms, financial institutions, and education agencies. For the data industry chain it is the same without exception.

3.2.1   Economic Entities

In the real world, a single enterprise can play all the economic entities' roles in the data industry: the data resource supplier, multilevel data product provider, and data product terminal customer, despite those roles are totally different. For example, Google's[1] dominance of the Internet is all of a data resource supplier with a 15.9% annual growth in data storage (2012 data on year-on-year basis), a data product provider occupying 72.1% market share (search engine), and a data product terminal customer that has US$1.71 billion (2011) of annual revenue from advertising placement.

1. *Data Resource Suppliers.*   In being resource oriented, data resource suppliers are indispensable in the data industry chain, and in general, they own huge accumulations of data resources. Although the data undoubtedly depends on the magnetic storage media, data resource suppliers do not just save data resources into their data centers, like the banking "safe deposit boxes." The data resource suppliers are obliged to convert data (e.g., express cleanup, partial clean or complete clean, erase privacy), store data, and provide paid services for downstream enterprises.

2. *Multilevel Data Product Providers.*   Customization can decide the diversity of the data product, and then the multiple levels of data product providers, including general level, secondary level, tertiary level, and so on and on. However, such economic entities may not provide data products together with their independent intellectual property rights (IIPR). This is because a data product that may be merely a content update in the business of a firm is based on the firm's historical data, such as workflow optimization, business development, marketing adjustment, and risk management, all of which may be covered in the original IIPR. Furthermore there are some major differences between data products due to different "sources," that is, data products processed from different data resources cannot interchange with each other.

3. *Data Product Terminal Customers.*   Increasing profits is a typical feature of the data industry and may induce the data industry to expand their markets. Data product terminal customers come from all of the present socioeconomic industries or sectors, and these industries or sectors will strengthen their own key businesses and increase their core competitiveness using customized data products.

3.2.2   Environmental Elements

Environmental elements also are vitally important in fostering and developing the data industry chain. Here, we chose the following six as examples for illustration.

1. *Institutes for Data Technology and Its Industrialization.*   To boost the data industry - a strategic emerging industry, we have to develop new bright spots of

---

[1]Conclusion of Pingdom and iResearch from 2012 worldwide Internet data monitoring & research.

growth, in which enhancing closer the industry/university relations is a pretty good option. One way to foster this relationship is through industry/university cooperation across the arts and sciences, that is, by industry getting affiliated with a university supported by industrial groups. Another is to plan on evolving into vital leader, such as, a major consulting group of providing data industry-oriented services for government, an important national repository of economic strategy data resources, an influential innovator of data technology and its industrialization, and a famous educational establishment of training data scientists and other talents. Studies in interdisciplinary methodology at academic institutions could include industrial layout of the data industry, data resources' reserve management, data technologies' applicable law and regulations, data products' circulation transactions.

2. *Data Industrial Bases.*   A data Industrial base, or a civilian industrial base of data enterprises and ancillary units engaged in related work, is essentially the "data technologies"-oriented Science and Research Park. Data industrial bases can attract highly skilled employees, especially if the office infrastructures with its multiple facilities for implementing industrial policy are located in. Data industrial bases for the layout must be unique, flexible, and clear enough, to adapt industry changes and, through the strengthened social ties, to promote cooperation and joint innovation with neighboring institutions or industries.

3. *Data industrial Investment Funds.*   Raising funds is often a big challenge in emerging industries, but the key factor is timeliness. Timeliness in acquiring capital is often a decisive factor in realizing a new data technology product. There are three time periods where fundraising is central to the success of the enterprise: the development cost in the initial incubation period, the revolving funds in the market expansion period, and working capital in the accelerated IPO[2] period.

4. *Government Departments.*   Government policies and grants from government agencies can strengthen and protect intellectual property in regulating the marketplace, providing tax concessions for innovative projects, and assisting in building industrial alliances or public service platforms, which can then further facilitate the formation, development, and evolution of innovative products.

5. *Intermediaries.*   Intermediaries include industrial alliances, market research companies, and business consultation organizations. All can be effective in providing communication platforms and contacts with "spokespersons" for the interests of data resource suppliers and multilevel data product providers. Market research companies and business consulting organizations can be good sources of information on related industrial developments.

6. *Public Service Platforms.*   There are three key public service platforms related to the data industry: data product pricing and valuation, data product circulation and transaction, and the training of talented workers in data technology and industrial skills.

---

[2]IPO: Initial Public Offering.

3.3   INDUSTRIAL CHAIN FORMATION

Industrial chain formation research focuses on value-added activities with a chain-structured base; and its fundamental motivation is to improve the value of industrial systems, beginning with value analysis and dimensional matching.

3.3.1   Value Analysis

There are at least two approaches in value analysis: one is to analyze the value-added attributes for the enterprise of the internal data industry chain; the other is to quantify the value of different industrial links. Both are critical at the industrial development stage. Some emerging industrial enterprises experience friction within the existing economic system. Such and many other problems need to be attended to, especially if the new enterprise desires to cover all economic activities in the entire industrial chain. A value analysis would therefore assist the enterprise in analyzing its value-added attributes from an internal operations perspective. Then, as an industry reaches a certain scale with its development prospects, the enterprise may consider implementing a subdivision or outsourcing services. But we will study the value of quantifying industrial links from an external perspective.

Value-Added Attributes Analysis   Suppose that a start-up enterprise desires to cover all economic activities in an industrial chain. The data industry chain has three value-added attributes: the data resource, data technology, and data product, as described below.

For the first value-added attribute – the data resource – no matter what the data resource is, the original value is already undoubtedly huge due to the continual creation of industrial subdivisions in the previous and existing economic systems. In such a case it may be difficult to increase the added value. Google is a good example. Google cannot gather data in another domain such as finance despite its having absolute dominance of the Internet. Second is the data technology. The original value or the added value of data technology is hard to measure because data technologies serve as tools in the production of data products. We therefore treat the data product as the most value-added attribute in the data industry chain.

At the present time the industrial chain of the enterprise under our consideration includes only one company base. The enterprise has yet to more or less attract and absorb some SMEs. When it does, it has two types of adsorption to choose from. One is the SME that has itself taken the initiative to acquire or outsource; for example, Facebook did a partial acquisition of Threadsy in 2012, and in this case the active entity was Facebook. The other is the SME that becomes voluntarily affiliated; for example, many search engine optimization firms appear to meet client needs that advertise on the front page of Google natural search, and in in this case Google remains passive or hidden. The result of the absorption is a merging of data resources, a transfer of data technologies, or an upgrade of data products, any one of which adds value to the data industry chain.

The data product is usually not completed all at once but is created stage by stage over a series of workstations from different enterprises. The data product, accordingly, has two characteristics: the continuity of the added value and the final co-creation value.

Therefore the enterprises in the data industry chain are of two types: leading and alliances. The leading enterprise and its alliances along the data industry chain establish the logical or space-time relationships for the purpose of effectively increasing the added value and maximizing profits. These industry leaders, with or without intention, compress the industrial chain and assume the vast majority of the added value. Their numerous alliances share the remaining added value from the disarticulated links. In the final value, if a co-creation, the leading enterprise and alliance enterprises make respective contributions based on their positions in the data industry chain. These contributions could refer to integrating data resources, developing data technologies, or improving the quality and value of data products, according to different conditions.

Difference Analysis in Quantifying Value of Industrial Links   The data industry chain is composed of nine industrial links: data acquisition, data storage, data management, data processing, data mining, data analysis, data presentation, data product pricing & valuation, and data product trading. Some industrial links (data acquisition, data storage, data management, etc.) also belong to the information industry chain, in a close-knit relationship.

It is generally accepted that value abundance of different industrial links is decided by their respective value-added spaces. For example, an industrial link could be rationed more funding because it is at the higher end of the industrial chain, and so has a larger value-add space. However, this is not true in every case.

1. *Data Acquisition.*   Data acquisition (aka data capture) belongs to both the information and data industry chains, and thus has a larger value-add space, which emerged due to Clintonomics (a portmanteau of Clinton and economics), whose mission was to implement a national information base to promote data acquisition (e.g., the 1993 National Information Infrastructure: Agenda for Action, the 1998 Internet Tax Freedom Act). However, because data is not only the output link of the information industry chain but also the input link of the data industry chain, the data acquisition link is no longer at the high end, despite having a larger value-add space.

2. *Data Storage.*   Data storage is also an industrial link that belongs to both the information and data industry chains. Firms specializing in the data storage link, such as the Internet data center (IDC), have commonly held storage media or devices. Undeniably, in the past, their single "store" business was the only way to make money. However, as a link having a certain amount of value-added space, the "store" business is not good enough to maximize the benefits and thus has to be upgraded. One way to upgrade the data storage business may be to construct "new" types of data centers, with environmental, electrical, mechanical, and networking improvements. In this regard Google is a forerunner. Google has a new patent on a water-based data center that includes a floating platform-mounted computer data center that has multiple computing units, a sea-based electrical generator connected all the computing units, and

one or more seawater cooling units for providing cooling to all the computing units.[3] Another way may be to enlarge the capacity of the storage media
to include advanced data storage techniques. For example, Sriram Kosuri of
Harvard University developed synthesized DNA as a target for data storage,[4]
and Nick Goldman of the European Bioinformatics Institute designed an information encoded[5] palm-sized artificial DNA capable of holding as large as 3ZB
of data. Yet, another way may be to improve the existing business model using
data innovations. As previously mentioned, this could be done by converting
data (e.g., convert data type, erase privacy) and providing paid data resources
to meet customer demand.

3. *Data Management.* Data management has experienced three stages of
   development: as manual management, as a file system, and as a database
   management system (DBMS). The primary purpose of data management is
   effective data organization, such as describing internal relationships between
   data, reducing data redundancy, or logically storing scattered data in the same
   tablespace. Second is to ensure effectiveness of the data, such as by searching
   required data using a SELECT statement. When gathering data into different
   versions of database management systems designed for narrow applications,
   data heterogeneity and complexity may present a big challenge. Currently,
   the three key data technologies – data processing, data mining, and data
   presentation – are gradually shaking off the influence of data organization. So
   the future development of data management is rather limited to the low end of
   the data industry chain.

4. *Data Processing.* In data processing there are two phases. Pre-processing
   refers to converting unprocessed data into a format suitable for analysis or
   mining (also called data preparation); post-processing guarantees the integration of the only effective and useful results for data analysis, data mining, and
   data presentation. As one of the three key data technologies, data processing
   is a tedious operation with different strategies and technical procedures (i.e.,
   aggregation, sampling, dimensionality reduction, feature subset selection, feature creation, discretizaton and binarization, and variable transformation [6]).
   Data processing also requires the practitioners to have sensitivity toward the
   data. For example, consider the dimensionality, sparsity, and resolution from
   data sets in pre-processing phase, converting categorical text data according
   to the attribute type of discrete to continuous numeric data (e.g., an eye color
   attribute compassing the notion of "amber, blue, brown, gray, green, hazel, red,
   and violet" be represented equally well by the values $\{1, 2, 3, 4, 5, 6, 7, 8\}$)
   in order to make the data suitable for mining. For another example, there is
   the post-processing phase, pairing k-means cluster results with another algorithm (e.g., single-link) in order to further define nonconvex clusters.[6] The data

[3]Google wrote in the patent US7525207.
[4]http://www.nature.com/nbt/journal/v28/n12/full/nbt.1716.html.
[5]http://www.nature.com/nature/journal/v494/n7435/full/nature11875.html.
[6]Convex sets: In Euclidean space, an object is convex if for every pair of points within the object, every
point on the straight line segment that joins them is also within the object.

processing link is indeed a key link in data industry chain and has potentially abundantly high value that right now is far from sufficient.

5. *Data Mining.* Through continually undergoing innovation, data mining is the most important industrial link in the data industry. Data mining also provides a key supporting role for other industrial links in finding hidden patterns and trend that cannot be detected by traditional analytic tools, human observation or intuition. For this reason, the data mining link is at the high end of the chain.

6. *Data Analysis.* There are two kinds of analytic thinking. One is to find appropriate arguments supporting a contention, and the other is to reveal a contention from the listed evidence. The traditional statistical approach is based on the first kind of analytic thinking using a hypothesize-and-test paradigm; specifically, the first step is to propose a hypothesis, next is to gather the data following a well-designed experiment, and finally to analyze data with respect to the hypothesis. Unfortunately, this process is laborious and does not really "let the data speak for themselves." Although data has no such capability, the second kind of analytic thinking often does not require an especially carefully designed sample to find or abstract a point in compiled results. It should be noted that the second one needs results to be evaluated by experts in the field. Therefore, although the data analysis is not a key data technology, this industrial link still plays a definitive role in the data industry chain.

7. *Data Presentation.* Data presentation mainly refers to visualization and visual effects. The development of the data presentation technologies has been motivated by the desire to help audiences members, on the one hand, quickly absorb the visualized information and, on the other , to be effectively stimulated to a new level of understanding by the material being presented visually. There are many data presentation graphics like stem and leaf plots, histograms, and pie charts, and which is used often depends on the nature of the data being presented (e.g., Chernoff faces are a technology for visualizing higher dimensional data, the contour plots are for visualizing spatiotemporal data). Many studies have been conducted on visualized data pre- and post-processing. In general, pre-processing can reveal some easily ignored aspects of data, and post-processing can help in understanding the results. However, there is little research on how to visualize data mining processing. Just as Yixin Chen, a data mining expert at Washington University in St. Louis, has noted, "Visualizing data mining process is very important for users to understand, interpret, trust, and make use of the results." The data presentation link combines key data technologies and innovations; it runs through the entire data industry chain and has great value potential.

8. *Data Product Pricing & Valuation.* Pricing and valuation is the only industrial link in the data industry chain directed toward public welfare. As a public service platform it may be useful for nonprofits in providing third-party evaluation, registration, and authentication based on market demand and international practices. However, here, a missing or faulty link can directly or indirectly influence the development of the data industry.

9. *Data Product Trading.* Data product trading can be used to track securities, futures, or electronic money transactions, and effectively all transactions involving the trading of data. This link covers multiple components, including registration, transactions, and settlement. Take a 1% transaction fee, for example. If each data product values US$1,000 at the trading rate of daily 1,000 items, then the daily transaction fee would be US$10 thousand. This link therefore has fantastic value potential, and is at the high end of the data industry chain.

In summary, among the nine links of the data industry chain, those with the most concentrated value are data processing, data mining, data presentation, and data product trading. But data product pricing & valuation as a public welfare industrial link are vital to a national data product evaluation, registration, and authentication system.

3.3.2   Dimensional Matching

Dimensional matching can have relevance to a study of the industrial chain's "docking mechanism" and pertain to all four dimensions: a value chain, an enterprise chain, a supply chain, and a space chain.

1. *Matching the Value Chain.* Matching an industrial chain for the value chain is a macroscopic "line-line" connection, that is, a two-dimensional relationship. This is important to the formation of the industrial chain and to its development and industrial value. In the data industry chain, the matching is dictated by the data resource. For example, Alibaba Group grabs its transaction data resources through its three platforms: Alibaba (B2B), Taobao (C2C), and Tmall (B2C). It has acquired a dominant e-commerce market position in China, and thus controls the survival and development of other enterprises in that level of the industrial chain.

2. *Matching the Enterprise Chain.* Matching an industrial chain for the enterprise chain is a madhyamaka "dot-line" connection, which is the relationship between an entity and its dimension. It can be considered a starting point of the industry chain from an enterprise's concrete form of embodied carrier relationships such as: enterprise vs. consumer, enterprise vs. enterprise, and enterprise vs. environmental element.

3. *Matching the Supply Chain.* Matching an industrial chain for the supply chain is a microcosmic "dot-dot" connection, which is the inside dimension of the relationship, indicating the manufacturing process and industrial level. Within is the division of industrial levels and technology, and specifically a production factors chain, a logistics chain, a demand chain, a product technology chain, and a technical services chain. In the data industry chain, this matching refers to matching the supply of data resources with provisions of the multilevel data products.

4. *Matching the Space Chain.* Matching an industrial chain for the supply chain is a "line-surface" connection, whereby related problems are divided into two

broad categories. One is to consider the distribution of similar types of industrial chain links in different regions, namely industrial distribution, matching
radius, and regional economic. The other is to consider the distribution of the
entire industrial chain according to geographic size, namely the global chain,
national chain, and regional chain.

However, matching the space chain can introduce a time matching problem. For
example, a matching of enterprises in the Western and Eastern Hemispheres can introduce uninterrupted 24-hour operations around the globe, and thus greatly increase the
operational efficiency in both hemispheres.

3.4   EVOLUTION OF INDUSTRIAL CHAIN

The preceding discussion of the data industry chain evolution started from "data
resource" oriented data technology development and ended with capital accumulation, as shown in Figure 3.1.

As compared to the development of other industry chains, it may be seen in the
figure that the data industry chain has approximately three stages. (1) At the primary
stage, the data industry chain is controlled by a single enterprise, which covers all economic activities in the entire industry chain. In matching the upstream, midstream,
and downstream links of the low degree chain, there might be discovered broken or
isolated links where the data technology is backward due to insufficient capital. (2)
At the intermediate stage, the data industry chain is basically complete; the data technology is under development with relative abundance in capital. (3) At the advanced
stage, the data industry chain evolves in an orderly way; the data technology is inter-
actively developed with active and vigorous capital. Therefore corresponding to the
evolution of the data industry chain are three additional evolutionary considerations:
driving factors, development pathways, and evolutionary mechanisms.

1. *Driving Factors.*   Based on Porter's competitive theory, and on evolutionary
   economics and industrial organization theories, an industrial chain responds
   to several driving factors, key among these are choosing a favorable competitive environment, breaking down the barriers to technological progress, and
   targeting a proper market selection by satisfying consumer demand. Besides
   applying these driving factors in the data industry chain, the practical application is essential to find opportunities to get social investment and government
   support.
2. *Development Pathways.*   There are four possible development pathways in
   the data industry chain. (1) Extending upward, mean that the information data
   industry can be "reset" for remodeling, since the data industry evolved from a
   reversal, derivation, and upgrading of the information industry with increasing
   profits from the properties of other industries. Undoubtedly, this is a win-win
   situation. (2) Exploring downward, is in fact limited because there are only
   two industrial links in the downward data industry chain – data product pricing & valuation, and data product trading. Data product pricing & valuation

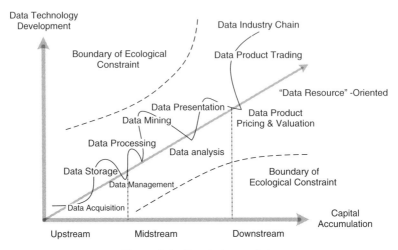

Figure 3.1 Data industry chain

is indispensable, but it is a public welfare link with a small value space. Data product trading is at the high end, but is subject to multiple constraints from data product quality, value, and related markets. (3) Vertical enlarging, is easily realized through the development of the data technology. (4) Lateral subdivision, can be controlled by both the data resource and data products market segmentation. Thus the evolution of the data industry chain can take the form shown in Figure 3.2.

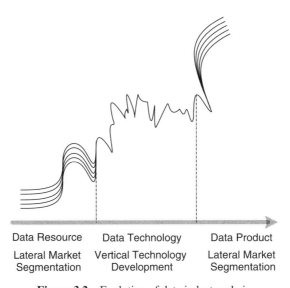

Figure 3.2 Evolution of data industry chain

3. *Development Mechanisms.* The development mechanisms of different industrial chains have many similarities as well as differences. Because the data industry chain is a resource-oriented industrial chain, its development mechanisms may start from how to reduce data resource dependence and focus on balancing between the competing interests of different industrial links.

3.5 INDUSTRIAL CHAIN GOVERNANCE

Industrial chain governance depends on the organizational relationships or institutional arrangements among the vertical-transactional economic entities of selected links from the relative upstream and downstream enterprises. The governance of the data industry chain is in effect an issue related to the coordination of different economic activities along the industrial chain, and this function can be classified into two categories. One is administrative in nature and pertains to internal governance; that is, coordinating internal units of the enterprise with internal economic activities that enhance core competitiveness. The other is external governance; that is, strengthening data resource cooperation, data technology R&D, and data product manufacture among alliance enterprises.

3.5.1 Governance Patterns

The data industry chain must meet certain governance conditions at different stages of operation. This section compares five governance patterns described by Gary Gereffi and colleagues [39] from Duke University (i.e., market, modular, relational, captive, and hierarchy governance).

1. *Market Governance.* Market governance can be accomplished through a series of arm's-length market relationships, among which market price is the connecting tie. The main feature of this pattern is maintaining transactions of low complexity while attending to high product standardization. For data products, it is rare for upstream enterprises to develop completely standardized products based on their own estimations of market demand; therefore market governance is not feasible when a data industry chain has not yet been established.
2. *Modular Governance.* Relative to market governance, modular governance is suitable for products with more complicated specifications, standards, and transactions based on a division of labor; in essence, it is a kind of contract manufacturing, in which the contract is the connecting tie. The main features of this pattern are highly complex transactions and high product standardization. For data products, the division of labor is in product demand and design (instead of conventional design and production) because data products may be designed specifically for a certain data resource. Modular governance is feasible when data industry chain has not yet been established.

3. *Relational Governance.*   Relational governance is suitable for interdependent transactions, and is normally achieved through reputation, social, spatial closeness, and family or cultural assurances in which "checks and balances" are the connecting tie (i.e., upstream and downstream enterprises are complementary in capabilities, with binding effects). The features of this pattern are highly complex transactions and low product standardization. Relational governance is a relatively ideal pattern for the future data industry chain, since during the data product manufacture both sides may only have an inkling of cooperation at the initial stage without a clear picture of the product profile and final results. However, once both sides put cooperative effort into products, each may well benefit in different ways.

4. *Captive Governance.*   Captive governance is suitable for exclusive transactions led by one of the upstream or downstream enterprises when an industrial chain has not yet been completely established or is in a disordered state, in which control is the connecting tie. The features of this pattern are low transaction complexity and low product standardization. Captive governance is an effective pattern for the data industry chain at the present stage, and the leading enterprise is the ultimate controlling party.

5. *Hierarchy Governance.*   Hierarchy governance is the only internal pattern among the five types, and is an extreme pattern of vertical industry chain governance; its organizational structure may appear as integrative or as a hierarchy, initiated by a department or link in the upper level that controls power resources, in which management is the connecting tie. Hierarchy governance is a typical pattern in the current data industry chain.

It should be reiterated that governance patterns in an industry change at different stages the same as the market changes or technology advances. In practice, therefore, the mixture of "giving hierarchy style of management the highest priority, with captive governance as a supplement" is recommended for the current data industry chain; whereas in the future, expansion in the dot-plane approach may replace top-down to be the relatively ideal pattern of relational governance.

3.5.2   Instruments of Governance

Governance patterns may be a way to rationally judge the relationships between the economic enterprises. The essence of efficient governance is in selecting or designing appropriate instruments of governance. Besides the institutional framework involving market price, contract, checks and balances, control and management processes (i.e., initiation, negotiation, supervision, revision, implementation, and termination), instruments of governance may include authentication of qualifications and relevant standards.

Institutional Framework   Creating an institutional framework is critical for economic transactions relative to upstream and downstream expenditures, namely to prevent opportunistic behavior. There are the legalities of contracts among these parties, articles of agreement pertaining to duties of the parties, penalties for breach of contract, costs and benefits computations, and, of course, arbitration issues that depend on this framework and allow cooperating parties to reach a common vision and avoid violations.

For the data industry chain there are a number of possible ways to build an institutional framework. (1) Regarding technology, besides other general technical parameters, indicator sets, and performance-testing methods that assess business (e.g., the information industry chain), specialized data is required for data products. (2) Regarding service, besides the scope, rank, and methods mutually agreed upon by cooperating sides, temporary services or a description of pricing principles based on changing demand should be included. (3) Regarding business, besides agreement on various ranked bonuses or penalties for meeting or violating commitments, an incentive policy for exceeding the agreed level should be included. (4) Regarding quality, besides data product quality evaluations or inspections, a summary of quality states related to the technology, services, and business sections, as well as measures that prevent slipups, should be included.

Authentication   Authentication covers data resources, data technologies, and data products. The objective here is to ensure that the asset value of the data resource controlled by the enterprise or organization, the level of data technology used, the technical level of its data products, or management level of the provided data services conforms to international or national standards, technical codes, or other mandatory requirements, including assessment of comprehensive conditions, level of performance, capacity of management, technical skills, and talent strengths.

Relevant Standards   Industry standards are the core consideration in administrating the data industry chain. Industry standards, together with an institutional framework and authentication, should be utilized by all participating entities. Let us briefly examine each economic entity or environmental element in the data industry chain. The data resource supplier has the responsibility to evaluate data resources and data assets. The multilevel data product provider is charged with upgrading the data product, as needed, and thus to safeguard the trustworthiness of the data services. The data product terminal customer is depended on to confirm demand for the product. The data technology and industrialization research organization is critical in providing instruction on related research, R&D, and to set up meetings on developments in data science. The data industry base instructs staff on building standards to meet specific functions. The data industrial investment funds are used to control capital input and recovery, guarantee profits, and reduce withdrawal costs. The government liaison department provides updates and instructions on regulations and developments in the data industry. Then there are the intermediaries used to coordinate the relationships among the alliance enterprises. Last, the public relations unit announces to the media key advancements in the enterprise services.

At present, relevant standards in the information industry applicable to the data industry include (1) system standards that cover to enterprise services, software capability, and advancements in internal procedures, such as the Capability Maturity Model (CMM) and Capability Maturity Model Integration (CMMI); (2) information safety standards that cover IP protection and information safety, such as the

Statement on Auditing Standard 70 (SAS 70); (3) information technology services standards that cover instructions on implementing any standardization upgrades, such as the Information Technology Service Standard (ITSS). However, presently the standards specific to the data industry are outdated and in dire need of re-design and modification.

3.6   THE DATA INDUSTRY CHAIN AND ITS INNOVATION NETWORK

The innovation spectrum of the data industry chain is relatively broad. The innovations can range from an incremental change to an essential overhaul, and so are of different scales. The innovation may be to the industry model, the transmission mechanism, the business model, technology, the product line, customer service, staff specialization, staff management, organizational structure, or market intervention. Innovation in the data industry chain can increase profits not only in an industrial link but for the entire industrial chain, and transform or upgrade the data product's terminal customers, raise the competitive advantage of region's economy, and then on upward and become embedded in a nation's reputation for industry innovation.

3.6.1   Innovation Layers

Unlike other industrial chains where innovation occurs at a single industrial link or between several industrial links, innovation in the data industry chain occurs across industrial chains (i.e., in the data industry chain with another industrial chain), though other layers may require different types of innovations.

1. *Innovation in a Single Industrial Link.*   An upgrade in a single industrial link may call for an upgrade by the data resource supplier and data product provider as well. Such independent upgrades may be done in different ways. For example, in the product pricing strategy, the existing one-time charge on data resources may be changed to data product piece pricing, or as another example, in internal operating procedure, the hard measures used to evaluate effectiveness may be dynamically optimized by soft measures set according to a team's behavioral preferences.

2. *Innovation between Industrial Links.*   The agents of innovation among the industrial links would involve all economic entities and environmental elements along the data industry chain. A change made to the governance patterns underlie its governance structure would require different types instruments of governance, specifically including capital operation mode, employment maturity model, symbiotic synergistic mode, and other selected management practices.

3. *Innovation across Industrial Chains.*   Innovations across an industrial chain may include mutual recognition of standards, collaborative innovation, and resource sharing. This layer of innovation may be due to (1) active expansion,

whereby the data industry chain has a more important role in that it expands its innovation network worldwide by inviting global enterprises or organizations with unique data innovative abilities to join forces; (2) passive absorption, whereby innovations result from forming a different industrial strategic alliance network by using existing expertise to attract customers from different domains to realize a transformation or upgrading.

3.6.2 A Support System

Each innovative support system of the data industry chain that has its own interest area needs its own economic entities and environmental elements to construct a powerful social network. For example, one measure may be whether economic entities are capable of effectively utilizing the knowledge spillover to obtain expert staff; another may be to evaluate whether environmental elements can actively contribute internally or externally to innovation in the data industry chain (e.g., providing timely capital input by data industrial investment funds), including government offerings of policy support or tax concessions, lending or preferential credit by financial institutions, and integrating resources due to intermediaries or public service platforms.

Human Resources Support   Human resources support (HRS) serves directly to recognize leadership and creativity of individual employees, and indirectly to enhance team practices and skill acquisitions by making use of the knowledge spillover effect for the purpose of adding value to human capital. These measures include but are not limited to (1) fostering an innovation culture, since innovation depends on staff's enthusiasm and creativity; (2) encouraging participation, such as adding an innovation category in a performance review, and reasonably arranging the work schedule of key innovation staff; (3) paying for training at external institutions.

Capital Support   Funding is basic to continued innovation in the data industry chain. Funding decides the direction, scale, and amplitude of industrial innovation, and thus the future industrial development, and as a result relatively low-growth industrial links get scraped due to a lack of funds. Generally, innovative activities exist only in links with sufficient funding support. Self-raised funds, government funding, and venture capital investments are the three sources of capital support. However, the funds raised by SMEs are limited by the utilitarian nature of government financial support, and avoidance of noncommercial R&D in research institutes required by commercial capital, and this restricts productive capacity expansion in emerging industries. A countermeasure may be to increase investment in a high value-added industrial link. Only when a large quantity of funds flow to the high value-added link will a unique investment structure be formed by key investments in a country or region in a given time period.

Public Service or Law Support   Public service should be a responsibility of the data industry base and intermediaries and be supported by government. The obvious benefit of public service platforms is to strengthen ties and cooperation within enterprises and organizations in order to raise the efficiency of the data industry in sharing and spreading innovation and effectively integrating data resources in political, academic, and business circles. At the same time, the competitive nature of such an environment makes it necessary for government to pass relevant laws and regulations that protect data resources and prevent oligarchic and monopolistic control of data assets in order to ensure open access to data but also to ensure data privacy and to prosecute fraud.

4

EXISTING DATA INNOVATIONS

There is an ongoing debate as to whether data innovation is disruptive or incremental. The "disruptive" side argues that the micro innovations aimed at improvement in the user experience are just the typical characteristics of disruptive innovations, as constant dripping water wearing holes in the stone, thus making many more small-but-excellent and fast-and-accurate enterprises sufficient to transform industry worldwide. The "incremental" side believes that the word "disruptive" is derogatory, and because breakthroughs are made little by little, the innovations should be incremental. No matter which side wins, it is undoubtedly that the power of innovation is rising.

4.1 WEB CREATIONS

The Internet has emerged as a global industrial medium that now drives web creations of all sorts. It is both a true outlet for self-expression and original work and for unique design conceptions and product customizations. In essence, the Internet has demonstrated the richness of grassroots' creativity beyond the orthodoxy of the mainstream.

It should be noted that the web creations, as intellectual achievements in the form of "data" from the first day of existence, are a type of primary data product that is continuously being innovated.

The Data Industry: The Business and Economics of Information and Big Data, First Edition. Chunlei Tang.
© 2016 John Wiley & Sons, Inc. Published 2016 by John Wiley & Sons, Inc.

4.1.1 Network Writing

In this section, we discuss network writing. We take the China experience as the main example here, given the unexpected rise and thrive of Chinese online literature.

Early Chinese online novels were entirely free. The novels were mostly full-length and posted on the campus forums of China's best universities.[1] The stories were mainly fantasies associated with the age and creative environment of the authors, who were mostly college students. These students' stories resonated with online readers and the return on such writing was not only long comments, which were limited by minimum number or length, but also renewed the stories depending on the comments. Some authors even spun their active readers' names into the novels of that time. This is an interactive creation between the authors and readers, and the readers often derived more pleasure in participating in the writing than the reading. *Injustice Ghost Road* written by Tinadannis is one of the more famous novels of that time. It developed an online following through serialization on Yat-Sen Channel in 2000 and soon became so popular that hit major popular public forums as the Lotus Seedpod Ghost Stories of Tianya.cn, conservatively developing an estimation of over thirty thousand reproductions on other websites.

From this point on, people who usually followed authors of original literature turned their attention to online literature for noncommercial purposes. A series of websites dedicated to this original literature were created, including Rongshuxiao.com, which advocated "a literature oriented toward the masses"; Jjwxc.net, which won the first prize in the original literary website's Monternet mobile PK contest; and Qidian.com, which was developed by the Chinese Magic Fantasy Union.

Network Writing's first data innovation employed a "bidirectional payments" business idea of Bookoo.com.cn (Bookoo), a Chinese writing and reading website established in the Silicon Valley, introducing "payment to authors under contract" and "payment to download and read." Despite having declared bankruptcy in 2001 due to the general "free to access" rule then in effect, Bookoo led the first commercial surge of the online literature. In October 2003, Qidian.com redesigned Bookoo's idea into a new "electronic publishing" fee arrangement that compensates the authors in one of two possible ways after signing agreements: a 50/50 equitable divided payment at a rate of 0.03 RMB per thousand Chinese characters, or an outright buyout of 50 to100 RMB per thousand characters. The fee arrangement brought at once a hot pursuit, and by 2011 the "web author" ranked among the "New Dreaming Careers' Top 10" released by Baidu.com. In June 2012, Tianqiao Chen, who had co-founded Shanda Interactive Entertainment, announced that Qidian.com had started to make a profit. By then, there were many more creative themes for Network Writing, such as fantasy, space-time travel, alternate history, officialdom, self-cultivation, armored robots, tomb, fortune teller, rebirth, and supernatural stories. Some themes did overlap in fiction, such as the self-cultivation novel including rebirth after space-time travel. The popularity of these online literatures led to a lot of new terms, such as "pit" (which

---

[1] These BBSs include BDWM BBS of Peking University (http://bbs.pku.edu.cn), SMTH BBS of Tsinghua University (http://newsmth.net), and Yat-Sen Channel of Sun Yat-sen University.

is an uncompleted original serialized work), "chasing" (refers to reading the work simultaneously with the speed of the author's updating), and "fattened" (means waiting for a period of serialization, then reading). But a sword is always double-edged. This innovation encountered a number of issues, such as authors forming an author team to write a novel in order to get much more pay, with millions of characters in length at every turn, or serious stylistic differences; then again, the pitiful fee very often causes authors to breach their contracts for various reasons, claiming illnesses, pregnancies, and even an excuse of "buying biscuits."

Network Writing's next data innovation was to use some data analysis technologies to market their original literature platforms. They tried to take the easy route, subdividing novels according to type, content, and writing style to allow readers to choose conveniently, for example, among romance stories, fairy tales, fantasy fiction, martial arts novels, or swashbuckler legends; or among stories of bygone times filled with revenge, nobility, and historical lives in accordance with content; or among works of drama, comedy, and tragedy. Then, for a story promoted on the homepage, based on readers' election results done by clicking, buying, tipping, and voting, they would upgrade author's contract type and guide author's writing in that direction. Other innovations included optimizing site search engines and generating personalized recommendations.

Today, after acquiring the leading seven original literature websites, including Rongshuxia.com, Jjwxc.net, Qidian.com, Hongxiu.com, Xs8.cn, Readnovel.com, and Xxsy.net, Shanda Literature, owned by Shanda Interactive Entertainment, controls nearly 90% of Chinese online literature market share.

4.1.2   Creative Designs

Creative designers often try to present unusual ideas, and their design activity involves planning, imagining, and problem solving in visual form. Designers work in diverse fields such as advertising, architecture, art museums, handicrafts, high fashion, magazine and book publications, software, music, performance arts, broadcasting, and filmmaking. Design services can be classified into graphic and animation design at three levels: preliminary design, secondary design, and creative design. Creative design is at the top level and that requires intense design preparation and the integration of concepts that not only demonstrates the designers' originality but also is sensitive to audience response.

On the demand side, designers must make their presentations satisfy a broad range of customer tastes. That is, a successful product is often based on its designer's artistic capability, creative style, innovative spirit, and sensitivity toward demand, market awareness, timeliness, and design pricing. Nowadays, many designers choose to list their information (e.g., personal profile, previous works, range of design work, client evaluations) on personal homepages or third-party platforms. Therefore some websites have added screening functions to help customers select a designer, such as classifying designers according to sex, age group, client rating, creative category, and cost of design, to shorten the search time, simplify the procedure, and reduce the unnecessary part of the workload.

On the designing side, a critical issue for designers is choosing what kind of creative materials of use, that is, from among the material data in various stocks (or royalty-free) of photos, images, footage, video, and clips libraries. There are three existing methods that help designers choose materials. One is a text-search by keywords or serial numbers that are marked on the material. Another is an image-search, namely a search for a picture through pictures. Third is an advanced site's search engine, referred to as Google's PageRank algorithm, which records the screening operations of designers, including input content, browsed addresses, and dwell time (i.e., to analyze whether the seventh material on page one, or the first material on page seven, was clicked) and then to change results and move up/down the ranking of the material according to the click count.

4.1.3   Bespoke Style

The term "bespoke" (meaning "discussed in advance") first appeared on Savile Road in the heart of London where men's clothing was tailor-made to the individual buyer's bodily measurements.[2] Today, more than ever in the past, it appears that more vendors are allowing customers to directly intervene in the production process and to demand a customized product, and this threatens to break down century-long standardization practices. Such a bespoke revival on the Internet is different from that of the agricultural or handicraft age when it might only be minor changes to a standardized commodity, or small modifications to the manufacturing procedure (e.g., printing a certain image or saying on a T-shirt does not change the original production procedure but does add a personalized touch). Sometimes the manufacturer may not be directly responsive to a consumer's demand but adopt partly a customer's comments for the purpose of providing personalized products with appropriate prices; for example, Volkswagen in China once launched a project called "The People's Car" and used an online platform to widely collect users' opinions and ideas as to the automobile's shape and design so that a perfect car model could be built. The project was mocked by some people, but to their surprise, in only one year, it attracted 33 million website visitors resulting in the submission of 119,000 unique ideas relating to automotive products and technologies.[3]

In the present-day bespoke movement, data is collected using survey questionnaires and cumulative historical consumption data to learn and analyze customer needs and preferences, and thus produce a standard commodity with a "personalized" design. So far, the existing methods have yielded two kinds of benefits. On the one hand, a consumer can get the equivalent of directing a manufacturer to produce an individualized commodity, or even to control the product's price; on the other hand, manufacturers can avoid not only the potentially wrong decisions due to narrow market research, and then to make a reasonable production plan, but also increase their brand equity and reduce the possibility of a unmarketable product.

[2]http://en.wikipedia.org/wiki/Bespoke_tailoring.
[3]http://www.shareable.net/blog/what-a-hover-car-says-about-the-future-of-crowdsourcing.

4.1.4 Crowdsourcing

Traditionally, project management involved such activities as planning, organizing, commanding, coordinating, controlling, and evaluating, which were implemented by a project leader under limited resource conditions in order to achieve a stated project objective.

In 2006, Feff Howe, who is a contributing editor at *Wired Magazine*, coined a term "crowdsourcing"[4] to describe a method of using the Internet to elicit help from any number of people willing collaborate on solving a standing problem. The idea came to Howe and his colleague Mark Robinson as year earlier as an upgraded version of outsourcing. Crowdsourcing is different in that a project leader only needs to dismantle the stated project objective into discrete units at the initial stage, and merge the project results at the end.

In 2007 a company called Clarizen was launched that delivers on-demand project management using crowdsourcing software in approximately the following content areas: (1) selecting a style of management that fits a project (e.g., "agile management"); (2) assisting a project leader in dismantling a stated objective into visual subprojects or subcomponents; (3) arranging suitable venues or dates to facilitate effective teamwork; (4) tracking a project's progress, and keeping the project leader informed of any project alterations or next milestone tasks; (5) automatically creating project analysis reports in order to reduce the frequency of meetings, emails, or status update reports. By virtue of its unique services, Clarizen raised US$35 million in a venture funding round led by Goldman Sachs & Co in May 2014, and brought in a total of nearly US$90 million. Wrike is a firm doing similar work; it has also attracted venture capital, and won US$15 million in the equity financing round led by Scale Venture Partners in May 2015.

Project crowdsourcing is not limited to above mentioned data innovations; it also can ease employees' workloads, which in combination with much needed rest can result in clearer thinking when faced with analyzing controlled resources, tracing project progress, and arranging long-distance real-time cooperation.

4.2 DATA MARKETING

A decade ago, in the golden age of relational databases, Arthur Hughes from Maryland University put forward the concept of "database marketing [40]" as a branch of direct marketing. Database marketing provides an enterprise with a continual collection of data for analyses related to customer satisfaction in order to deliver to the customer higher value; this was to cultivate customer loyalty, and thus to establish and retain "customer resources."

However, "data marketing" is not equivalent to "database marketing." Unlike the highly accurate data of database marketing as based on a relational database system, data marketing can only integrate heterogeneous and complex data: it is a marketing method based on mining massive data, which includes database marketing. Although

---

[4]http://crowdsourcing.typepad.com/about.html.

less popular now, because of the ability to handle richer data, data marketing did breakthrough the market barriers in collecting a huge number of applications.

4.2.1   Market Positioning

Market positioning, also known as marketing positioning, is one of the three core factors in marketing. According to positioning theory advocates Jack Trout and Al Ries, positioning starts from the moment a product is launched and then is modified based on existing competitive products. Only modified are the "name, price, and packaging," not the substance. The existing product has to be made to look different so as to establish a brand, and build up the enterprise's core competitiveness. Hence the key to market positioning is in the product analysis process.

There are already innovations in data marketing concerning product analyses that aim at the market positioning core, including analyses of product differentiation, target audience, and market feedback. Product differentiation is a key factor in the structure of the marketplace. This is true universally except in a perfectly competitive market (i.e., product homogeneity) or an oligopoly market (i.e., single type of product). Therefore the extent of control an enterprise has over a market depends on its products' "significant" differentiations. From a marketing perspective, "significant" differentiation refers to how well in tune the homogeneous product is with consumers' purchases and preferences. Take airfare sale, for example. The generic means is to analyze historical order data, learning about the time interval and price class of advanced reservations, and the approximate price differences between direct flights and connecting flights, and to identify the customers' decision processes, so as to adjust the discount level dynamically.

The target audience analysis provides a comprehensive way to consider unknown customers' interests, behaviors, and performances, in order to aggregate different types of customers and trace their consumption paths. More specifically, by this process the key characteristics can be abstracted of a certain unknown customer group, and labels attributed to that group, in order to realize the positioned divisions of the target group. Another example is the online shopping cart, which monitors mouse clicks, grouping the customers who repeatedly add and cancel the same merchandise, and then recommend them similar commodities with lower prices or higher quality.

Market feedback is an important inspection link in the PDCA (plan–do–check–act or plan–do–check–adjust) of marketing system. It is used to generate and report marketing results, based on the degree of product differentiation, and to adjudge the decision-making and implement a marketing strategy. In this link, data innovation has broken through the traditional approach of acquiring data via survey questionnaires, analyzing nested models, and using data tools like Zillabyte and GoodData to mine multi-source complex data, and finally to track market status.

4.2.2   Business Insights

Business insights fall into the scope of third-party marketing. Third-party marketing is a consulting service that aims at the improvement and innovation on the original marketing system with the purpose of achieving added value, which can be typically divided into (1) active cooperation, whereby the manufacturer takes the initiative to cooperate with third-parties and share the revenue in an agreed-upon proportion, or (2) passive acquiescence, whereby the manufacturer tacitly allows a third-party to join marketing activities and realize a mutual win. Business insights belong to the latter method.

Business insights start with market surveys (including industrial sectional views) and lead to analyses, predictions, or case studies in multiple parts on brand confirmations, advertising promotions, and performance examinations, in the hope of finding business opportunities in a vast market in order to accelerate the brand building, make rational strategic decisions, and promote marketing efficiency. From this perspective, business insights are related to data collection and analysis.

A brand is not only a trademark; it concerns reputation, products, corporate culture, and the overall operations management. What's more, Philip Kotler, a global expert on marketing, described the brand as a "symbol of the pride, passion and enterprise of its members and represents creative works." Compared with maintenance and improvement of a self-owned brand by manufacturers, brand confirmation from third-party marketers is relatively objective, so that can help a manufacturer clearly understand market feedback in order to improve product differentiation. Brand confirmation normally includes (1) brand distribution analysis, which involves penetration rate, possessing capacity, and market attention; (2) dimensions of brand competitive analysis, including product quality, market performance, and transaction processing; (3) brand sales potential forecasting. At present, there are multiple third-party marketers centralized in the processing of complex data in many aspects, among which are brand identification (e.g., high reputation, innovation capability, market influence, product leadership), channel construction, customer expansion, media presentation, searching ability, marketing activity, and public praise as a type of brand recognition. The most prominent are the World Brand Value Lab's BSI (Brand Strength Index) and Baidu's Brand Probe.

Advertising and promotion are critical to marketing; and the criterion in judging the effects is whether the target audience is identified by their media habits. This is because conceptually behind advertising, storyboard tests, and assessments of advertising effectiveness are nested at each stage a target consumer group. Lacking of such awareness will directly lead to a failure of all work done, and especially, lacking knowledge of the media habits of consumers can be highly detrimental to an advertising campaign. Today, targeted advertising is in vogue; it has displaced a previous marketing concept that emphasized "content or theme." Because the Internet is now the chief medium, targeted advertising has higher pertinence in identifying the interests or preferences of target users.

Performance examination is somewhat similar to market feedback. It is the inspection carried out by a third-party marketer for the purpose of accessing key information about the entire sector or various levels of a domain, to identify returns on investment of promotional activities in real time and business performance management, and to make supply chain predictions on customer acquisitions (e.g., customer loyalty). In this regard innovative data solutions normally implement synergic analyses on large-scale Internet data, especially where there is a large quantity of supply and demand information, on historical performance, and on current trends of products, and financial statements disclosed by public companies.

4.2.3 Customer Evaluation

Data on customer behavior is important for current customer management assessments. Customers are generally assessed, from a marketing perspective, by how they fit within three categories: actual customers, possible customers, and competitors' customers. An actual customer is one whose demand has been satisfied by the rights and powers of purchase (products or services), and likely has had several satisfactory transactions with the company. The other two types are often grouped together as potential customers.

Marketing systems that use data to determine customer buying habits can be roughly divided into three categories too: traditional marketing, database marketing, and data marketing. Due to customers' complex and volatile purchasing behaviors and manufacturers' limited resources, the first two types focus on actual customers in order to control client resources and prevent loss of their customers. Compared with today's improved traditional marketing that has adopted an assessment scale, database marketing is based on a relational system that can create complete and accurate profiles of customers, and implement in-depth detailed anatomic understanding of these customers, so as to better learn a single customer's preferences based on the history of that person's web activities. However, database marketing also has some drawbacks: (1) it was hard to analyze potential customers, and (2) it was not possible to pigeonhole buying habits, preferences, interests, and near- and medium-term demands of actual customers.

Data marketing makes up for these drawbacks and, by mining large-scale unstructured Internet data, can (1) generate sales patterns that connect consumers' purchasing behavior toward different brands and product features; (2) compare the products purchased or services patronized by consumers (e.g., quantity, brand, type) with potential repeat behavior; (3) analyze consumer cyclical purchasing habits or brand switching to find contextual factors influencing purchases, preferences, and buying habits of customers; (4) obtain different profiles of consumers to study market segmentation based on multiple contextual factors.

Typical data tools include Mintigo and Consumer Portrait. Mintigo can analyze product features or specific market information keyboarded by manufacturers; the process allocates a specific "code" called CustomerDNA as an important clue, and then searches the web, social networks, and third-party databases to find prospects that fit the potential customer profile. Using keywords through customer subdivision, market status, and section analysis, the potential consumer is profiled to determine interest points, network media contacts, regional behavioral differences, life styles, and brand perceptions.

There are a variety of consumer profiles that belong to companies. Here, we take Baidu's case study as an example. In 2011 Baidu and Procter & Gamble (P&G) launched a prototype Olay product. Baidu found the new Olay product to confuse loyal customers. P&G made adjustments accordingly, and then put on the market a similar Olay product tagged to be suitable for 25-year-old women; this turned out to be a product easily accepted by the market.

4.3   PUSH SERVICES

Push services are based on expressed preferences for information, and a server push technology that "describes a popular style of Internet-based communication where the request for a given transaction is initiated by the central server."[5]

Because the B/S (browser/server) mode can provides cross-platform, maintenance-free client services, as well as extendibility to a multiple client platforms and a friendlier interface than the C/S (client/server) mode, more web applications have begun to switch to the B/S structure. However, the B/S mode has a defect in AJAX (Asynchronous JavaScript and XML) applications: there is no persistent connection and bidirectional data transmission between the user's browser and a server. The browser therefore initiates a new long polling request in order to simulate real-time communication (e.g., updating a stock price). This method not only wastes server resources but also produces a delay every time it establishes or closes a connection. For example, the process of the iOS push mechanism that is an application server – apple push notification service (APNs) – searches iPhone via pop-up notification. So long as APNs and the iPhone can communicate smoothly, all updates can reach users immediately and cut data flow to a minimum; otherwise, the data synchronization that results in such frequent delays can sometimes become unbearable.

To reduce data synchronization, some service providers often analyze users' behaviors (e.g., favorite, forward, ignore, and block operations from different sources or themes) before applying push services, to screen out users with higher hit ratios. There are roughly three types of data innovation: targeted advertising, breaking news, and message broadcasting, all of which have the features of fragmentation, diversification, and real time.

4.3.1   Targeted Advertising

The term "targeted advertising" originates from a data product – Baidu's targeted advertising – released by Baidu at the Baidu World Conference in 2006. Baidu even applied for a patent on this product. It works in a simple way that combines the "ranking bid" with online advertisement delivery. Once the potential customer is locked in, advertisements are placed and traced on multiple super packet data traffic channels (PDTCHs) at an effective frequency until clicking occurs and applicable fees are paid.

As the use of targeted advertising has expanded, its meaning has shifted from the general term that refers to an advertisement push mode that traces and analyzes cookie files of the user's browser to a more specific function that locks in a particular group or customer according to the sponsor's desired outcome. Targeted advertising

---

[5]http://en.wikipedia.org/wiki/Push_technology.

has developed rapidly from the initial stage of geographical area and the second stage of customer interest, to the current stage of user behavior. Gmail, for instance, only pushed advertisements by matching the keywords in the mail body text to the side bar of the page at its earlier stage; later it further analyzed the senders' characteristics (e.g., whether an email has been read or not); and currently, in addition to advertisements, it pushes group purchases and discount reminders according to a user's geographical area through Google Offers.

Various portal websites are also keen on targeted advertising, and the transformation began with the financial crisis of 2007/2008. This was when the big three sector clients were faced with huge financial problems: the automobile entered a low-growth stage, most real estate sectors experienced capital chain ruptures, and the IT industries were busy self-rescuing themselves with layoffs. Since advertising revenues declined badly, these portal websites had to adjust their business focus from "packaged-sales" news to targeted advertising, with which they became overjoyed. The same fixed advertising space was used to deliver advertisement against different users, which in the end gained much higher revenue from the fees they charged multiple sponsors.

Another form of targeted advertising is real-time bidding. This is an innovative business strategy that combines both targeted advertising and price comparison, and involves media and digital advertising agencies. Real-time bidding is used to buy and sell advertising inventory instantaneously based on an ad impression,[6] selected advertisers from among many ads, via a programmatic instantaneous auction that matches buyers with sellers. More specifically, a transaction begins with a user triggering a bid request that may include various pieces of data, such as user's information, browsing history, and location. The request goes from the publisher to an exchange-related page. With bidding that often happen within 100 milliseconds, buyers can only bid on an impression; that is, the bid is won, and the buyer's ad is instantly displayed on the publisher's sites. Real-time bidding can help avoid the conventional issues of repeated crowd coverage and waste of advertising resources.

4.3.2   Instant Broadcasting

Broadcasting is the distribution of content (e.g., text, images, audio, video, multimedia) to a dispersed audience. Today, broadcasting can be done by any electronic mass communications medium[7] in a one-to-many model. In the past, broadcasting services, typically only used the electromagnetic spectrum (radio waves). In the last decade, the Internet as emerged as an alternative via IP broadband, mobile, fixed and satellite networks. Accordingly, we can distinguish the former and latter as analogue and digital broadcasting. Compared with analogue broadcasting, digital broadcasting has an instant push feature despite some small time differences. The user's browser can receive content almost simultaneously, even when a client (e.g., mobile phones, tablets, laptops) is powered off; in turn, updates such as news and personal messages are instantly pushed to browsers, streaming fastest the generated contents.

---

[6]http://en.wikipedia.org/wiki/Real-time_bidding.
[7]http://en.wikipedia.org/wiki/Broadcasting.

There is one such example of push notifications concerning breaking news from *Sohu News* during the China (Ya'an) earthquake in 2013. The first push service was on April 20 at 8:15 am, only 13 minutes after the outbreak of the earthquake; about three minutes later (8:18 am), the Live News Room was online, and at the same time relative search and safety reporting programs were added, broadcasting from the disaster area was live for the first time; by 10:19 pm, the total participants had reached 2.23 million. Of course, not all news broadcasting is instant in form, but the 2013 earthquake changed the one-way dissemination of traditional news and made reporting interactive, enabling eyewitnesses to report events as they occur and thus interact immediately by electronic media.

Messaging itself shows a lot of differences. The differences are sometimes reflected in (1) the message content, such as position services often push positioning or map updates, games push invitations or chat, and video apps push the recent hot topics; or in (2) the data type, including text, images, audio, video, and multimedia. Users, in general, have high expectations for the broadcast push message. Recent research from Localytics[8] shows that sending highly targeted, personalized content resulting in three times higher conversion rates. So there should be more data innovations for message broadcasting to motivate interactions between publishers and users; for example, to differentiate user types through data mining on background, activity levels, and relationship chains, or to compare the login profile with the login device's current position to make sure that messages are 100% delivered.

In summary, instant broadcasting that relies on high quality can get "sticky" users; the content of a data product has to show innovation. Currently, besides news and messages, many other types of contents are being pushed, including online courses, travel trips, and healthcare tips.

4.4   PRICE COMPARISON

The term "price comparison" is not the same as "price parity," which is an economic concept that describes the relationship between merchandise price and major currency at the macro level. Price comparison means a simple comparison of price based on factors such as merchandise value (or transformed form of value) and supply-demand relationship.

A price comparison tool is also a primary data product, and its popularity is due to e-commerce, which made price comparison profitable. To filter and compare products aggregated from many different online retailers involving price, features, and other criteria, price comparison tools came into existence and were developed based on the optimization of search engines. Merchandise data in these tools usually comes from the Internet and is obtained by a web crawler; some may be obtained by sharing data among cooperating parties. In addition the interactive trading mode represented by O2O (online-to-offline) has created a mixed online/offline price comparison tool that collects merchandise information using barcode scanners or QR code (quick response code) readers and guides consumers to "first search, then compare, and finally buy."

---

[8]http://info.localytics.com/blog/52-percent-of-users-enable-push-messaging.

At its current stage, data innovations are occurring in this field in developing price comparison derivatives and implementing dynamic pricing.

Price comparison derivatives refer to content innovation extending to append goods' background knowledge, forgery prevention queries, and drawback tips. The goods' background knowledge includes functional description, use specification, consumption guide, typical applications with common sense, as well as shopping skills and fashion trends. Forgery prevention queries typically concern authoritative forgery prevention descriptions and various quality inspection reports related to the reliability of a commodity, which can hamper fraudulent or fake marketing. Drawback tips prompt relevant national or industrial standards for personal and property safety, for example, the detection of exposure to dicyandiamide from a milk importer of New Zealand. Apart from these, one-stop shipment tracking is a special price comparison derivative involving evaluation of delivery services; it is an innovation in data resource sharing. One-stop tracking that calls the data interfaces of different express companies is more convenient than traditional queries via the official website. It not only can track the location of parcels but can implement courier service comparisons (e.g., delivery of scope, service charges, complaints) to enhance customer satisfaction in the logistics link of e-commerce.

Dynamic pricing is a series of standard pricing strategies in the airline industry, including time-based pricing, market segmentation, and limited supply strategy. Because of the lack of physical stores' geographical price advantage, e-commerce retailers take dynamic pricing more as a common way of activating inventory, ensuring shipment, and enhancing purchase rate. Existing data innovation in dynamic pricing is to forecast price of goods based on price comparison. In Jane 2011 Decide.com[9] gained US\$8.5 million from Maveron, Madrona, and other angel investors for the idea of "price predication and recommendation," and has raised a total of US\$17 million outside funding to date. Decide's mode (1) solves price tracker and comparison along the timeline, (2) carries out data mining through several dedicated algorithms involving millions of price volatilities and their influencing factors (e.g., new product release cycles, news reports, company bulletins), and (3) presents with low-threshold operational experiences and full visualized results. Since October 2012 Decide has offered price forecasting services that are guaranteed to give compensation for any forecasting failure, changing the previously free forecasting into a monthly charge of US\$5 by subscribers or US\$29.99 with guaranteed compensation. In this program, if the price of a commodity keeps declining after the best buying time given by Decide, Decide will pay the price difference as compensation. Through this strategy, Decide has saved US\$127 million of potential spending for over 10 million users.

4.5   DISEASE PREVENTION

Disease prevention is actually a wonderful prospect for humanity, but that takes integrated multiple factors as the object of study, including disease, pathogenic factors,

[9]http://techcrunch.com/2013/03/22/decide-com-raises-8m-series-c-for-its-price-predictions-engine-will-expand-to-more-categories-improve-data-mining-mobile-apps.

organisms, crowds, and society. Disease prevention can in fact be divided into four levels: prevention, control, elimination, and eradication.

Mythological stories abound, like "Shennong tasting a hundred herbs,"[10] and have spread over thousands of years to prove the unremitting efforts in conquering diseases by humankind. Today, we depend on clinical and experimental medicine to intervene and effectively control diseases. In recent years, accompanied by the explosive accumulation of data in the medical field, increasingly more numbers of clinical experiments are becoming data experiments, with data as both experimental subject and experimental object. So data innovations have now materialized in the area of disease prevention.

4.5.1   Tracking Epidemics

Generally, when the observed count of a certain disease exceeds the expected value, we call it an epidemic. As an important component of preventive medicine, epidemiology studies the distribution rule and influencing factors of population health and disease. Traditional methods design questionnaires for surveys, conduct observations and inquiries, collect data and establish statistical models, to examine the factors related to certain diseases.

Search engines as preliminary data products allow for the identification, tracking, and forecasting epidemics. A milestone in this regard occurred in November 2008, when Feremy Ginsbert and colleagues [41] at Google published an article titled "Detecting Influenza Epidemics using Search Engine Query Data" in *Nature*. Google and Yahoo tracked and found the transmission pathway of influenza using, respectively, their search engines in the following the virus for two years. In 2010, Google again released "Google Flu Trends" based on search results to order to give an early warning.

Other innovative methods in tracking and forecasting epidemics exist today. In the 2012 International Conference on Digital Disease Detection, Nicholas Christakis, who is Professor of Medical Sociology at Harvard Medical School and author of *Connected: the Surprising Power of Our Social Networks and How They Shape Our Lives*, conceived of a way to trace epidemics by the use of all the social networking sites. Sickweather, a Boston start-up honored among "100 Brilliant Companies" (2012) by *Entrepreneur Magazine*, started implementing this vision and to use social media as Facebook and Twitter to follow outbreaks of the flu, allergies, and other illnesses. Sickweather can alert users every time they are in the vicinity of a sick person, for example, before you enter a shop or sit on a crowded city bus, you will know immediately if someone inside has had a fever in the past 24 hours or been exposed to a chickenpox-ridden child at home.[11] Compared with Google's Flu Trends "48-hour lag" and reports from CDC "several weeks behind," Sickweather has an advantage in real-time reporting.

---

[10]http://en.wikipedia.org/wiki/Shennong.
[11]http://www.mysanantonio.com/lifestyle/article/Alert-apps-user-sneezer-nearby-4888816.php.

4.5.2 Whole-Genome Sequencing

Since the 1950s, biological research has extended from the cellular level to the molecular level. In 1986, Renato Dulbecco, the Nobel Prize winner in physiology and medicine (1975), published an article [42] in *Science* proposing whole-genome sequencing.

The essence of whole-genome sequencing is integration and analyses of biologic data implemented by applying advanced data technologies. There is the case of the legendary CEO of Apple, Steve Jobs, who died of pancreatic cancer (a disease with high morbidity) in 2011. The standard survival time from onset of this disease to death is usually half a year to one year. How could Jobs live for eight additional years and create business miracles at Apple, even after an accurate diagnosis of his tumor? The result is, to large degree, attributable to his sequenced genome – Jobs is one of the only 20 people in the world who paid to have complete whole-genome sequencing. Multiple research organizations from worldwide top universities participated in the analysis of Jobs' entire genome data, including Harvard, Stanford, and Johns Hopkins University.

Ever since work on decoding the human DNA molecule began, many scientists have become fascinated by the DNA structure. As Stephen Meyer, director of the Center for Science and Culture at the Discovery Institute in Seattle, has said, the DNA stores an exquisite "language" composed of some three billion genetic letters.

World famous companies in the whole-genome sequencing industry include Incyte and DNA Direct in the United States, and deCODE in Iceland; then there are 23andMe and Editas Medicine. 23andMe, which was picked by *Time Magazine*'s Best Inventions of 2008 as truly expert at gene data analysis, using gene data mining from the medical history of ancient ancestors to learn genetic relations, searched the genome and identified the gene clusters that have large influence on certain traits. 23andMe also constructed a social network using similar genetic traits based on genes. Editas Medicine, a Cambridge, Massachusetts–based genome editing drug maker, can translate CRISPR/Cas9 and TALENs technologies into human therapeutics that enable corrective molecular modification to treat disease at the genetic level. By the way, both companies have worked closely together to Google. Google invested around US$10 million in 23andMe; meanwhile, Anne Wojcicki, the cofounder of 23andMe, is the former wife of Google's founder Sergey Brin. Editas Medicine, backed by Google, filed for its first IPO with an initial size of US$100 million in January 2016.

5

DATA SERVICES IN MULTIPLE DOMAINS

In this chapter we revisit the seven domain data resources mentioned in Chapter 2 and discuss some data innovations that may appear in the future in these specific domains.

5.1 SCIENTIFIC DATA SERVICES

For many years, scientific research was divided into three separate areas: basic science, applied science, and theoretical research, all depending on collected data that can support or reject a new perspective for reason of self-consistency, insight, accuracy, unity, or compatibility. The motivation was a quadrant research method- ology conceived to facilitate scientific development, and once popular in academic circles. The most famous was the four-quadrant configuration [43] conceived by a well-known development economist Vernon Ruttan, so called because it was based on Bohr's quadrant, Edison's quadrant, and Pasteur's quadrant, and used to understand the core characteristics of research (e.g., driving source, external restrictions, regression, availability) in advancing knowledge.

 However, the basic methods of discovery and the generation of data used in the past to acquire cutting-edge knowledge can no longer satisfy our needs in extrater- restrial explorations, in probing the inner Earth, in saving the environment or even humanity itself. Evolution had been an inevitable force in all pursuits of knowledge throughout the whole history of scientific exploration. Today, it is innovations that can accumulate data that will facilitate academic training, scholarly communication, and completion of scientific research projects in academic circles.

The Data Industry: The Business and Economics of Information and Big Data, First Edition. Chunlei Tang.
© 2016 John Wiley & Sons, Inc. Published 2016 by John Wiley & Sons, Inc.

5.1.1 Literature Retrieval Reform

To be an outstanding scholar requires strict and systematic academic training. There is, on the one hand, the need to cultivate the work ethic and research skills in use of appropriate references and, on the other, keeping up to date with the latest readings, discussions, speculations, and reviews of frontier literature. Setting aside the issue of academic learning for the moment, let us focus on the literature that enhances learning.

Differentiating and refining research categories (and even disciplines and specialties) can become an intractable problem, and this is a serious challenge in subject areas where the literature data is so massive and heterogeneous that there is insufficient annotation and even nonstandard terminologies used. A particular problem is that most downloadable versions are scanned images or photographs of literature, despite the fact that the much of that literature can be improved electronically. However, there is really no quick interoperable way to browse directories of academic literature. The current literature retrieval depends on a manual keyword search, and this can be very tedious and time-consuming.

Luckily, though the scope is still small, some literature data sets have been implemented using text mining tools and and methodologies including hypertext markup language (HTML), semantic web standards, and version control mechanisms. For example, "Reflect,"[1] a fully automated device, was developed by the European Molecular Biology Laboratory in Germany, and it operates as an external service on any webpage or as a browser plug-in to tag genes, proteins, and small molecules, and connects them to related external data entries. Another example, in 2009, David Shotton and colleagues [44] of Oxford University developed the CiTO (citation typing ontology) for the literature data on tropical diseases, and published it in *PLOS Computational Biology*. The CiTO can analyze "reference citations in scientific research articles and other scholarly works" as background information, known precedents, and related rhetorical work (hypotheses, claims, supporting statements, refutations, etc.), and also can do abstract statistics, rearrange references, connect to other research literature, and import data into Google Maps.

With the CiTO, we can imagine a future literature search process in which a literature data set includes semantic interoperability and, by means of an advanced search engine, can extract abstracts in return for a query. In addition to statistics and computing reported official impact factors, we could index literature according to peer review and get citation analysis diagrams; we could introduce visual engines into mathematical formulas or create algorithms for literature; we could even merge the latest works in certain research areas, and predict future research trends affecting information jobs.

5.1.2 An Alternative Scholarly Communication Initiative

Scholarly communication[2] refers to a sharing of research findings among academics and scholars. Usually the findings are communicated at seminars, workshops,

---

[1] http://www.elseviergrandchallenge.com.
[2] http://en.wikipedia.org/wiki/Scholarly_communication.

forums, conferences, symposia, and summits held at an academic institution. Sometimes these get-togethers can break down boundaries in scientific research or remove barriers among disciplines, when the assembled group includes a diverse scientific community.

Yet, the reality is that because of the discretion necessary in pioneering research and the growth of science teams, scholarly communication is faced with many problems, all at one time. These problems include (1) funds, whereby some researchers lack financial support to attend various academic meetings, or some academic meetings cannot raise enough money to invite the top experts; (2) reputation, whereby most scholars and researchers are aware of an intrinsic hierarchy in participating academic institutions and may disdain participating in academic activities at lower ranked universities or scientific institutes; (3) review system, whereby academic achievement is not closely connected to a peer-review system within the publishing system. As a result scholarly communication is often a disguised form of broadening academic contacts and capturing academic resources.

Future scholarly communication will likely be dedicated to re-examining research results and combining scientific records or citation data, instead of interminable meetings; for example, to understand the research route map of a particular author by repeating and summarizing his/her findings, or otherwise, to track cross-disciplinary research directions in several geographic areas by organizing scientific records or citation data around authorship.

To achieve this aim, we need to reform the ways in which scientific records are managed: (1) at the producing source, which is based on the open archives initiative protocol for metadata harvesting using multiple sources (e.g., semi-supervised learning, active learning) for listing citations in new scientific literature, in other words, call for a little additional effort by the author in exchange for the convenience of other scholars; (2) at the retrieving source, which has established machine operable expressions identified by a uniform resource identifier (URI) so that authors can upload their scientific records (e.g., the workflow process for research) simultaneously with the publication of their academic papers and other researchers can avoid overlapped work.

5.1.3   Scientific Research Project Services

Public investment in scientific research encourages academic competitions. Thus, obtaining a research grant, besides the high prestige and dollar amount, is an important criterion of measuring the strength of scientific research by the applicants whether this be an individual or an organization.

In general, research projects of all levels require an application process for the desired project funding, approval of the application, organization and implementation, review and evaluation, acceptance and assessment, declaration of achievements, promotion of achievements, and archiving, whereby at each stage a lot of paperwork is submitted. For example, there is an application proposal at the approval stage, project progress reports at the implementation stage, and a final report at the completion stage; other important original records include the approval document, job specification, letter of authorization, study design proposal, agreement, experimental study and survey, list of papers, a slides presentation, achievement award material, statement of income, and a list of expenditures. All these written materials are then digitized and fall into the category of the scientific record.

Throughout the years, international nongovernmental organizations such as the International Council for Science (ICSU) and its affiliates, the Committee on Data for Science and the Technology (CODATA) and the World Data System (WDS), have been promoting open access of scientific data sharing to include methodology, technology, standards, and management. The ICSU enjoys strong legal and policy support from many countries and regions, and it particularly favors international cooperative research projects.

Nevertheless, open access of the scientific records is not enough. Scientific research project services need data innovations to what now appears to be insufficient data because of (1) heavy paperwork and (2) insufficient data management. Take the application proposal as an example. Specifically, the applicant must include a project name, credentials, team composition, theoretical basis for the proposed research, an abstract, academic background, the research objective, significance to the field, outline of contents, methodology to be employed, a technical roadmap, a feasibility analysis, a results forecast, innovative points of interest, and a reasonable budget. While some categories are quite similar with minor differences, writing proposals can be time-consuming and quite prone to error.

For the funding agency, inefficiencies can arise at the point of monitoring the applications for duplication due to the diverging levels of research proposals (e.g., from vertically national and regional levels to horizontally enterprise and overseas opportunities). To prevent duplication of research among the multiple scientific research proposals submitted, the data management system should include a way to detect project overlaps, the same as a paper recheck.

For applicants, the data management of scientific research proposals not only should reduce the tedious workloads but also prevent abuse by unscrupulous applicants. Thus data innovation could be good for reviewers in detecting anomalies in multiple proposals and in identifying differences among applicants, besides relieving the workload of reviewers and maximizing fairness. The same may be applicable to industrialization fund initiatives that encourage SME development.

5.2   ADMINISTRATIVE DATA SERVICES

"Open government" initiatives as well as its "collaborative administration" initiatives are often highly praised in the Western academic circles as synergistic frameworks combining ideas, structures, and modes of operation to evaluate government performance in public administration. One of the typical cases is the Worldwide Governance Indicators (WGI) project of the World Bank updated annually since 2002, namely to promote voice and accountability, political stability and nonviolence, government responsibility, regulatory quality, rule of law, and control of corruption.

Over the past twenty years, the growth of electronic media has led to improvements in government administrative functions. However, compared with the actual public demand, the practicality and effectiveness of government services still leave plenty of room for improvement. In the main, there are still difficulties due to inappropriate authorizations, leaked secrets due to improper disclosure of information, and the like. Thus, in this era of e-government, implementing collaborative administrative data innovations is critical to real public demand for services, to ensure that public services are timely, accurate, and efficient.

5.2.1 Police Department

The police department is an administrative unit of government that is tasked with public security management, emergency management, and incident management. It is obvious that administrative data from police department has the property of strong timeliness. In October 2012 the world's top collaborative solution provider, SAP, invited a famous consulting firm, Penn Schoen Berland, to carry out a survey about real-time data in the public sector. The results showed[3] that among the nearly 200 respondent IT officers in the US government, 87% of federal government IT officers (100 people in total) and 75% of state government IT officers (98 people in total) thought real-time big data was extremely beneficial to a police department, as it can be used for crime prevention instead of settling a simple crime. For instance, a forecast of possible times and places of criminal activities could significantly reduce the overall crime rate in certain areas.

As a matter of fact, non-real-time criminal data is also useful. Many companies, including IBM,[4] have started crime forecast programs. In November 2011, IBM released an advertisement that demonstrates how to use big data for crime prevention, and alleged to have reduced the crime rate of some US cities (including Memphis, New York, and Charleston, South Carolina) by approximately 30%. In this advertisement, police officers rush to the scene before criminals commit a crime, just like a reality edition of the Hollywood movie *Minority Report*. For a similar purpose, a provider of security and identity solutions, Morphotrak, offers biometric identification systems for fingerprint (or palm) identification and facial recognition. ParAccel (which was acquired by Actian in April 2013) provides crime forecasts of high reference value to law enforcement agencies, through cooperation with SecureAlert, a criminal monitoring company. From this data, ParAccel obtains and analyzes over fifteen thousand individuals' whereabouts 24 hours a day. Moreover some research institutes are participating. Robert Lipton [45], of Michigan University, prepared a frequent crime hotspot map (2012) by combining multiple sources of data on alcohol sales, drug-related crimes, public security status, and population statistics in all districts of Boston with historical criminal records, in order to help the police find the most crime-prone city blocks.

Of course, police departments' administrative data is not the only type of crime data. Take China, for example. The official website of the Ministry of Public Security releases crime data (violent criminal and economic criminal investigations, along with detention criminal data), citizen ID data developed for population management, 110 call data vs. police recorded data, 119 call data vs. fire control

---

[3] http://fm.sap.com/data/UPLOAD/files/SAP_INFOGRAPHIC_BIG%20DATA.pdf.
[4] https://www-03.ibm.com/press/us/en/pressrelease/37985.wss.

data, biometric entry and exit data developed for border defense management, and vehicle registration data. In addition to this routine work, the Ministry monitors transportation data (e.g., civil aviation, railways, road transport) and cyber networks.

Yet, apart from forecasting and tracking crimes, data innovations in police departments still do concentrate on solving relatively small "social" problems. For example, simple response modes for small incidents (e.g., unlocking a forgotten-key door, moving a blocked vehicle, reporting a lost property in a taxi) that do not require dispatches of police force but rather direct notification of a locksmith, the car owner, and the taxi driver after verification of the incident. Additionally there are the processes that can be analysed dynamically by images collected from a monitoring camera on the crossroad, such as the interval length between alternating traffic signals based on real-time vehicle and pedestrian streams.

5.2.2   Statistical Office

The government statistical office, the original "social thermometer," has been criticized in recent years. Take the nonfarm payroll employment statistics from the US Bureau of Labor Statistics (BLS) as an example. In a report from October 5, 2012,[5] the household survey showed the drop of the unemployment rate from 8.1% in August to 7.8% in September, whereas the number of the unemployment decreased by 456,000. Oddly, the same report on the following pages illustrated graphically that the employment increased by 114,000 in September using another statistical approach. The statements were inconsistent and irreconcilable.

Inarguably, now is the time for statistical offices to consider improving their methods. The past shortcut to collecting, processing, analyzing, and interpreting "minimum" data in order to get "maximum" information should be replaced or supplemented by data technologies based on "all available data." Take the Consumer Price Index (CPI) as another example. The BLS always employs the use of survey questionnaires by way of telephone calls, mail, fax, or door-to-door visits; this method takes about US$250 million to collect prices for nearly eighty thousand types of commodities and the data lags behind the release of results. In contrast, the Billion Prices project[6] by Roberto Rigobon and Alberto Cavello, of the Sloan School at MIT, collects millions of global merchandise prices from hundreds of online retailers, in real-time and at high frequency; an inflation index can therefore be calculated instantly to determine asset prices or price increases. In September 2008 this project detected inflation and deflation of the US domestic trends about three months ahead of the official data release.

5.2.3   Environmental Protection Agency

According to a report[7] in *IT World*, a known dynamic IT information publisher, in May 2013 the US National Aeronautics and Space Administration (NASA)

---

[5] http://www.bls.gov/news.release/archives/empsit_10052012.pdf.
[6] http://bpp.mit.edu.
[7] http://www.itworld.com/article/2827431/it-management/new-google-timelapse-project-shows-how-earth-has-changed-over-28-years--video-.html.

and United States Geological Survey (USGS) requested that Google sift through over 909 TB of approximately two million images since 1984, translate them into time-lapsed video-like footages for the web to show delicate changes of landforms on this planet, including land reclamation in Dubai, deforestation in the Amazon, and a glacial recession in Colombia. The purpose was to document geological changes, natural disasters, and human activity, and thus to facilitate environmental work.

For the past century, the scope of human production and social activities has continued to expand, from the continent to offshore and ocean, and from surface to the Earth's interior and to outer space, in such magnitude as to have triggered a critical global ecological imbalance. Governments around the world are now more or less faced with overexploitation of nonrenewable resources, worsening pollution, and deterioration of the ecosystem. In order to address these issues, in January 1973, the United Nations General Assembly (UNGA) created the United Nations Environment Program to coordinate overall environmental protection planning the world over. UNGA is charged with appraising the world's environmental status and facilitating international cooperation on improving the environment. Many countries subsequently set up special environmental protection agencies, including the Environmental Protection Agency in the United States (established in 1970), the Environment Agency in Britain (established in 1996), and the Ministry of Environmental Protection in China (upgraded in 1998).

However, due to traditional data analysis techniques' limitations, environmental protection agencies have been basically monitoring the atmosphere, surface water, and radiation. Thus, despite coming into contact with data over a long period of time, the data analysis is still very weak. One important point, which should be noted, is that various environmental data sets are overlapping, no matter what type of related issue is addressed, including ozone depletion, atmosphere pollution (smog, acid rain, and toxic air), forest devastation, desertification, water or marine pollution, biodiversity decrease, and even climate change overall. The advanced approach is to list all available data, find the correlations among data, identify multiple genuine impact factors, and trace the source of various pollution chains in order to allow countries to adopt rational corrective measures.

5.3   INTERNET DATA SERVICES

While Internet data innovation is undoubtedly trivial, it is important to improve user experiences. Google is a well-documented example.

5.3.1   Open Source

On October 5, 1991, Linus Torvalds released Linux, which was developed as a free operating system using components provided by the GNU-project (GNU's Not Unix!) launched in September 1983 by Richard Stallman. Since then, open source has emerged following the principle of shared benefit and work. Red Hat, headquartered in Raleigh, North Carolina, became a precedent – "even if a non–Red Hat employee can obtain returns of a contribution" – which causes Red Hat rapidly on the list of "the top 10 Linux contributors".

Open source generally follows the model that each individual is responsible for partially related work and freely sharing the results with the others through communication, division of labor, and collaboration on a network. The open source collaboration is where the essence and core of data technology exists.

First, from the very beginning, the core technologies of data mining have been algorithms that are oriented to an application, including C 4.5, K-means, SVM, Apriori, EM, pagerank, Adaboost, KNN, naive Bayes, and Cart. The sharing of algorithms is in fact in accord with Copyright Law whose the legislative purpose is an idea–expression dichotomy, meaning "protection is given only to the expression of the idea – not the idea itself."[8] Algorithms belong to the idea, while software is an expression showing the nature of product. This is why Kingsoft's WPS and Microsoft's Office are similar without infringing on each other.

Second, most big data tools (e.g., HDFS, Hadoop MapReduce, HBase, Cassandra, Hive, Pig, Chukwa, Ambari, Zookeeper, Sqoop, Oozie, Mahout, Hcatalog.125) use Apache licenses that are issued by a nonprofit organization – the Apache Software Foundation (ASF, founded in 1999). Apache specifically supports the development of open source.

At present, open source is increasing "muscle" strength, and even Microsoft has compromised. A milestone was on March 26, 2014, when Microsoft finally published the source code for MS-DOS, one of its most important operating systems. In addition, according to Google's Android 2014 Year in Review[9] released in April 2015, "in 2014 less than 1% of all devices had potentially harmful applications installed," which is a proof that open source is secure.

5.3.2 Privacy Services

In September 2011, Forrester Research, an independent research firm, released a report titled *Personal Identity Management* in which was claimed[10] that third-party data on individuals' privacy trading in the US market totaled US$2 billion. In the report, analyst Fatemeh Khatibloo pointed out that "consumers are leaving an exponentially growing digital footprint across channels and media, and they are awakening to the fact that marketers use this data for financial gain." However, whether it is in the Anglo-American legal system or Continental law, privacy separated from property rights is an evolutionary direction of tort laws.

The 1890 essay "The Right to Privacy," appearing as a *Harvard Law Review* article by Samuel Warren and Louis Brandeis [46], is the standard by which privacy is defined, which is the "right to life" and "right to be let alone." This means that the liberty of any individual is up to his own willingness, in that an individual may choose to have liberty or give it up. Therefore consumers should have more control over the data products made from their private data, or, at least, the right to shared decision making and the right to know of the data usage.

---

[8]http://en.wikipedia.org/wiki/Idea%E2%80%93expression_divide.
[9]https://static.googleusercontent.com/media/source.android.com/zh-CN/us/devices/tech/security/reports/Google_Android_Security_2014_Report_Final.pdf.
[10]https://blog.personal.com/uploads/2011/10/Forrester-Research-personal_identity_management.pdf.

Currently, there are at least two privacy services offered consumers who are willing to "foot the bill": private data custody and privacy deletion.

Private Data Custody Private data of custody originates from ideas of crime prevention. Particularly after the "9/11" attacks, worldwide national security activities came to be more or less involved in individuals' private data. Some governments even established a series of laws and executive orders to increase data transparency and bind enterprises to store customer private data for long periods of time and to make regular reports on the certain contacts.

Some organizations began innovations on the custody of private data. OpenID[11] is a nonprofit international standardization organization that lets developers authenticate users "without taking on the responsibility of storing and managing password files"; namely a single digital identification allows users to "register at one site but login to multiple other sites." This is a replacement for the traditional "account plus password" authorization in a website and relieves users' memory burden. OpenID's current members include Google, Microsoft, Yahoo!, Symantec, Paypal, and PingIdentity.

Additionally, there is a start-up from 2012 at Harvard named SlideLife, with the slogan "Never repeat yourself." SlideLife provides a kind of "universal login" that accepts only a mobile number without filling any forms. Today, it can be applied in making purchases online, subscribing to paid services, checking into a hotel, registering at a hospital, and enrolling at a school.

Privacy Deletion Viktor Mayer-Schonberger's idea "to set up a storage period for private data" [47], suggested in his book *Delete: the Virtue of Forgetting in the Digital Age,* has won multiple awards[12] and was praised highly in academic circles. However, data has a time paradox [5], so why not try to only remove the privacy section?

In July 2011, a start-up called Reputation.com[13] raised US$41 million series D financing led by August Capital, and this brought the company's total capital to more than US$65 million. The company provides a new solution for individuals' privacy that is not simply deletion but can "suppress negative search results" and even deeply bury the data. The scope of work they handle includes blogs and website content, legal or financial information, newspaper articles, business postings, images, and video. Michael Fertik, Reputation's founder and CEO, describes their work as "selling" clothes to people "running naked" on the Internet. He revealed that the next step would be selling a "data antidote" to consumers who "claim their online profile and use it to travel anonymously through the web." Another company called Snapchat has started 2012 providing similar services with the unique application of "burning after reading." By March 2015, Snapchat was at a US$15 billion valuation and had rejected several acquisition offers from Google, Facebook, Tencent, and other giants.

---

[11] http://openid.net.
[12] *Delete: the Virtue of Forgetting in the Digital Age* was granted Marshall McLuhan Award (2010) by The Media Ecology Association and Don Price Award (2010) by American Political Science Association.
[13] http://blogs.wsj.com/digits/2011/07/18/reputation-com-raises-a-new-41-million-round.

5.3.3   People Search

"People and their relationships" are always attractive. The development of the Internet and social network provides a convenient channel for "people search," and it is reported that "30% of all Internet searches are people-related."[14]

People search is a vertical search engine on names based on matching algorithms on record linkages. There are great business opportunities in people search, and even nonprofit organizations are drawn to this feature. Finding information on a person of interest is a bright spot in people search, but the cunning part is individuals' relationships. Whereas early social studies in behavioral science very often focused on interpersonal relationships such as formation, structure, position, strength, and even the level of trust [48], empirical data has hit a bottleneck. Nowadays, Internet data is a substantial resource. For example, according to Pingdom.com,[15] Facebook may be the largest "country" without national boundaries on Earth by 2016 – "having 1.44 billion monthly active users"[16] – despite the fact that it is still blocked in China.

Innovations in people search using Internet data may have the following characteristics:

1. *Discovering Valuable Relationships.*   On the one hand, this is reflected in that before establishing contact, people sift through suitable objectives for similar backgrounds, common hobbies, or overlap of common friends, and, on the other hand, in that after establishing contact, they determine the value of the relationships and decide on the long-term value (or the value of different periods).

2. *Breaking Maintenance Limit of Relationships.*   Based on the previous research work, the weak ties limit has a value of 150, inferred by Robin Dunbar in 1992, and the strong ties limit has a value of 3, due to Nicholas Christakis in 2009. Breakthroughs are reflected in the maintenance of valuable weak ties and increases in valuable strong ties.

3. *Establishing Personal Networking.*   According to existing data technologies, personal networking can be abstracted into "points" and "lines" by adopting a graph theory proposed by Leonhard Euler, the forerunner of modern mathematics. Establishing personal networking can evolve into the construction of social reciprocal rings, based on multiple angles, such as the time, locations, and network extension directions, which present effective pathways between a user and the target user through description, measurement, and calculation of the relationships between the "points."

5.4   FINANCIAL DATA SERVICES

Financial data resources were the first to be commercialized; forecasting future price is in fact a common financial data service.

---

[14]http://techcrunch.com/2007/04/11/exclusive-screenshots-spocks-new-people-engine.
[15]http://royal.pingdom.com/2013/02/05/facebook-2016.
[16]http://www.statista.com/statistics/264810/number-of-monthly-active-facebook-users-worldwide.

In the past, it was mathematics that flexed its numerical muscles in the financial engineering. The task in financial engineering was to provide fast and accurate calculations grounded model-based applications, including the value-at-risk metrics, high-frequent or ultra-high-frequent data analyses, and pricing of derivatives. Today, we have a better alternative approach – data mining, which replaces the conventional hypothesize-and-test paradigm. For example, in sorting price fluctuations of various securities and commodities in different periods, data mining can help investors answer important investment issues that traditionally were too time-consuming to resolve, such as the tendency, form, price changes, and volumes, or the ever-fluctuating characteristics of financial products, and how to place a stop loss & profit target like a professional, in order to quantify the return required by an investment on the basis of the associated risk.

5.4.1 Describing Correlations

Modern financial markets endow finance – a discipline focusing on the law and judgment of value – with extensive reach. Both the macroscopic and microscopic aspects of financial relations have great research importance.

In studies involving financial relations, Stephen Ross, a well-known financial economist who developed the Arbitrage Pricing Theory, called for "intuitive judgments" [49] that include experiences, memories, and guesswork (e.g., the relationship between finance and news, the relations between financial development and economic growth). Ross argued that "theory is a handmaiden that attempts to bridge the gap between the intuitions and the data."[17] Taking an extreme view, he claims that financial intuitions have been largely diluted by hypothesis-based testing and empirical evidence. Likewise Merton Miller and Eugene Fama, Nobel laureates in economics in 1990 and 2013, respectively, are both supportive of financial intuitions. Admittedly, precise and complicated mathematical tools can easily get people lost in the different kinds of theories and models. Their translation and validation of financial intuitions may even mislead judgments. Therefore simpler and more intuitive tools are needed in studies of financial relationships.

Data mining tools generally are designed to implement a mainstream object-oriented technology that is interchanged with a structured methodology. Object technology is based on the concept of "object" and utilizes inter-object relationships to understand the connections between the existing objects and those corresponding to them. As a result different data mining tools manage, handle, mine, analyze, and demonstrate data and inter-data relationships in object space. All the relations (e.g., dependency, association, aggregation, composition, inheritance) are expressed not as causations but as correlations.

Correlation and causation imply forms of two confusable relationships between objective realities. Correlation is intended to quantify the mathematical relations between two objects, such as positively correlated, negatively correlated, and

---

[17]Original text is: The intuition and the theory of finance are coconspirators. The theory is less a formal mathematical structure driven by its own imperatives—although that is always a danger—as it is a handmaiden that attempts to bridge the gap between the intuitions and the data.

unrelated. On the basis of expressing correlations, causation further describes necessity and sufficiency between two objects, including one cause and one effect, multiple causes and one effect, one cause and multiple effects, and multiple causes and multiple effects. Since a causal relationship stems from a correlation, a human weakness occurs in that people have become so accustomed to look for causes in order to transform uncertainty into certainty in face of any type of uncertainty from the objective world. Such a psychological misunderstanding is frequently used in the real world, and a common trick is to utilize a real correlation to support an unproved causation.

Empirical analysis of hypothetical causation can therefore be replaced by mining massive data. The new mining tools as well as new ideas not only provide "intuitive" perception of correlation and certain connections that "haven't been noticed before" but also allow us to specifically describe and explain such correlations in order to increase the understanding of the social dynamics at both the microscopic and macroscopic levels.

5.4.2   Simulating Market-Makers' Behaviors

Modern financial theory (MFT), built on the capital asset pricing model (CAPM) and efficient-market hypothesis (EMH), has been continually questioned because of all sorts of abnormal phenomena occurring in the market ever since the 1980s. In 1979, influenced by behaviorism (also known as behavioral psychology), Daniel Kahneman and Amos Tversky jointly put forward expectancy theory, which is the first theory that can effectively be applied to explain systematically market anomalies, starting from actual decision making in applying mental processes to real choices regarding market behavior. Current practices in both domestic and overseas investments show that behavioral finance related theories and their applications in decision making provide a good perspective for understanding financial markets, a transition from "ought to be" to "under the real facts."

Unlike mainstream securities markets (encompasses equity markets, bond markets, and derivatives markets) dominated by institutional investors, China's securities markets is based on individuals. Therefore general methods to regulate institutional investors, such as warnings on simultaneous buying behavior and selling of certain stocks, are not suitable for controlling individual market-makers' speculative behaviors. In China, anyone with the slightest experience in securities investments knows that simulating behaviors of the market makers (i.e., try to fathom and study) can increase one's own bargaining chips of competition in "wealth redistribution." In a general way, the behaviors of a market maker can be divided into four types: position, pulled high, finishing, and shipment. The existing behavioral analysis methods only allow simple classifications of the market makers, for example, (1) operating practices: the academicism who has "a solid theoretical foundation and classic trader tips," and the pirate style featuring as "quick buy-in and forcible pull-high"; (2) latency time: short run, medium run, and long run; (3) duration of movements and amplitude: homeopathic buy-ins or gone against the tide, making profits or trapped in the market, and strong or weak. From this we cannot learn more about the time point, market size manipulation, and operating style for a layout for securities made by certain market makers. However, the market makers of ever-changing means have accidentally left clues in massive financial data, confirmed by similar fluctuations of different periods in certain securities that show similar behavioral patterns in the way a single market-maker's acts. Take stock market as an example, the market makers spread bad rumors at the beginning, then sacrifice a small amount of bargaining chips to suppress a stock in "excellent shape" (i.e., from growing), and thus let the market seem generally not optimistic. Later on, they carry out a homeopathic buy-in and use certain operating practices to make the stock price rebound, and finally take the opportunity to sell. As a result other investors are too late to follow them and bound to be trapped in the stock market.

Data mining is capable of automatically "identifying recurring features of a dynamic system" [6]; it can replace human visual inspection as applied in the field of behavioral finance for trend analysis, morphological analysis, and fluctuation analysis. For example, mining of large-scale financial data sets can detect interesting frequent patterns corresponding to operating practices of investors by excluding the presence of noise along the timeline and, by simulating makers' behavior, reveal manipulations of markets based on the real facts.

5.4.3   Forecasting Security Prices

In the financial field, spot trading and over-the-counter (OTC) trading depend on a change in value of the "underlying" asset. Such changes are measured through price prediction. Generally, forecasts for changing trends are more meaningful than accurate prices. Only by learning from the trend forecasting that utilizes fitting and other methods can we have a possible predictive value from the track points in the trends.

In forecasting studies there are many methods used: (1) modeling, which utilizes time series of long memory and is based on linear thinking, using expectation and variance of two angles; (2) trend analysis based on chaos theory and fractals of the random vs. determine, gradual changes vs. sudden changes, and disorder vs. order studies; (3) tracking evolutions in life movement characteristics of price changes, such as metabolism, profit-orientation, adaptability, plasticity, irritability, variability, and rhythmicity; and even (4) movements in price trendlines from parallel channels' angles of geometry, physics, and astronomy. However, all these studies are based on hypotheses. The Internet prophet Kevin Kelly unveiled the answer to why predictions are usually wrong in his book *The Technium*: "It's hard to transcend current assumptions."

Evolutionary data mining, one of data mining tools, supports research on forecasting without prior knowledge. For instance, we can apply a hyper-heuristic algorithm to create a series of random rules, and then use a similarity-based iterative or emergent method to describe the optimal price trend of the partial or entire objects over time.

Apart from methodology, the volume of data is another critical factor for forecasting. First, more volume of data can help increase knowledge. As described in *The Art of War* "if you know the enemy and know yourself, your victory will not stand in doubt; if you know Heaven and know Earth, you may make your victory complete." It is knowledge gained from big data that apparently goes beyond small data selected according to a well-designed random sampling plan. Second, bigger data volume allows for better data fault tolerance. Data fault tolerance is the property that allows the least possible error avoidance strategies to be used, or gets probable approximately correct results in the presence of deviations. When analyzing the large data sets obtained from multiple channels, "inaccuracy and imperfection may allow us to better understand the development trend" [1].

5.5   HEALTH DATA SERVICES

Medical and health sciences encompass the development of therapies, treatments, and health services, including clinical, medicinal, biomedical, behavioral, social sciences, and epidemiological research. Since this field is directly related to the health of all human beings, its commercial value has high earnings growth and high equity returns, and thus health data innovations are undoubtedly increasing.

Yet, compared to finance, which is virtually an area without experts trained in earning money, the healthcare field is filled with specialists, among which physicians, clinicians, pharmacists, and nurses are the better known. That is to say, in finance, it's not whether you're right or wrong that is important, but how much money you make when you're right and how much you lose when you're wrong; it is nothing like that in healthcare. A bottleneck to health data innovations is that practitioners find it too difficult to cope with data – they do not believe in the results of data mining the way financial managers do, but rather in their specific medical knowledge or experience. The fact is that Google's AlphaGo[18] (2016) does not require a great deal of prior knowledge to perform well; the innovation is not only in the methods and tools but in the ideas as well. It's time to take a look at health big data for potentially more varied applications, instead of just for text mining or genotyping alone.

Currently, Merck & Co. Inc. is working together with the Regenstrief Institute, on a five-year study (from November 2012), to share the LONIC database,[19] which aims to develop personalized medicine for chronic conditions such as diabetes, cardiovascular disease, and osteoporosis. The Blue Cross Blue Shield Association (BCBSA) is deploying a "continuous learning center" (CLC) with NantHealth, a big data supplier, in a bid to "deliver evidence-based care that is more coordinated and personalized."[20] Michael Thali of the University of Bern (Switzerland) has coined the term "virtopsy,"[21] a virtual autopsy system promoting the intervention of data experiments that, apart from finding the cause of death, may help medical teaching, evaluate nursing services, improve medical devices, and test military equipment.

---

[18]http://www.nature.com/news/google-ai-algorithm-masters-ancient-game-of-go-1.19234.
[19]LOINC: Logical Observation Identifiers Names and Codes, is a database and universal standard for identifying medical laboratory observations.
[20]https://www.blueshieldca.com/bsca/about-blue-shield/newsroom/nant-100212.sp.
[21]https://www.nlm.nih.gov/visibleproofs/galleries/technologies/virtopsy.html.

5.5.1 Approaching the Healthcare Singularity

After studying the lag between medical discovery and practice in the past 2500 years and in the past 150 years, Michael Gillam and colleagues of the Microsoft Medical Media Lab published a paper [23] arguing that the vast amount of health data have significantly reduced this lag, and that together with data technologies, it will provide a "singularity" – clinical instant practice, approaching around 2025.

Although it remains unknown whether or not 2025 is the specific year for healthcare "singularity," the conclusions of this study basically match with the facts. Take the example of the H7N9 bird flu outbreak in Shanghai, China, in late March 2013. The *Shanghai Mercury* reported that two American-owned biological companies, Greffex and Protein Sciences, announced the successful completion of H7N9 bird flu vaccine. This discovery pointed out the future direction of vaccine research and development, namely to find out the characteristics of virus through gene data analysis and to insert a heterologous gene sequences into an adenovirus carrier to develop a protein vaccine. The entire process only took one month.

Gilliam also described a post-"singularity" world. In one scenario, the protective measures would be instantaneously taken with a drug's withdrawal from the market for proven side effects: patients would be informed and the physician would immediately prescribe alternatives. In another scenario, medical residents could obtain personalized treatments as the pathogenic gene is discovered by genomic testing, and drug would be correlated with their genomic profile.

These scenarios are not impossible. For instance, if we want to understand the distribution of patient populations, based on existing experiences with data mining: the first step is to define similarity according to extracted multiple features such as geographical regional, genomic profile, and dietary habits (e.g., vegetable, meat), or addiction (e.g., smoking, alcohol); next is to cluster patients using their similarity feature, and then to analyze correlations between patients, so as to further subdivide patient clusters in order to provide personalized health services, such as dispensing different drugs to patients with similar diseases, according to their drug sensitivities, pathogenic genes, and so forth.

5.5.2 New Drug of Launching Shortcuts

According to the regulation standards of the US Food and Drug Administration (FDA), a drug is categorized as a new drug, generic drug, or over-the-counter drug. Because, worldwide, the pharmaceutical industry is connected to national welfare and a people's livelihood, countries typically use "new drug" as a sign of their national innovative capabilities. This is slightly different from definition of new drug in the world's major economies in which generally a new drug is defined by domestic approval, registration, manufacturing, and market size, as well as by new chemical substances, new compound preparations, new formulations, or new route of medication. Some countries (e.g., the United States) even include new indications, new specifications, new manufacturing locations, and new salts, acids, and other minerals, of the marketed drug into this definition.

Given the novelty of new drugs, any slight change to ingredients and dosage will affect its safety and effectiveness accreditation and certification. Therefore it generally takes a long time, about 9 to 20 years, for a new drug to come to market, over which time testing accounted for about 80%.

The risk for testing a new drug is obviously high. Thus the issue of drug testing frequently comes up. The World Medical Association (WMA) and the Council for International Organizations of Medical Sciences (CIOMS) separately instituted clauses for protecting drug test subjects in the Declaration of Helsinki and the International Ethical Guidelines for Biomedical Research Involving Human Subjects: in principle, "the potential subject must be informed of the right to refuse to participate in the study or to withdraw consent to participate at any time without reprisal."[22]

As early as in 2004, the Critical Path Institute (C-Path) noticed that sharing test data can expedite the clinical testing of new drugs. In March 2006, the C-Path partnered with the FDA, which acted as a consultant, and initiated a collaborative agreement[23] to set up the Predictive Safety Testing Consortium (PSTC) consisting of GlaxoSmithKline, Pfizer, Novartis, Bristol-Myers Squibb, Johnson & Johnson, Merck, Schering-Plough, and Roche, to share preclinical animal experiment data.

In the future, such data will significantly expedite human clinical testing of new drugs and reduce time to market. Data experiments will optimize the design of clinical trials. Clinical trials of new drugs will pertain to the empirical study of evaluating limited issues such as whether the patient gets better or suffers from side effects, instead of studying the functional mechanism of the drug. Therefore drug testing will proceed in two steps. First is to screen the test treatment group by matching or partitioning methods based on a patient's genomic sequence, biological tags, and electronic medical records. Second is to extract prior knowledge from previous clinical trial data of similar drugs cross-referencing the new test treatment groups' data. There will be included means to implement association analysis by connecting to other health data sources to track or manage testing results, for instance, if in a given period there are indications in the test treatment group of a new drug's metabolism, toxicology, or adverse effects.

5.5.3   Monitoring in Chronic Disease

Chronic disease is a generic term for diseases with latent start, long duration, and a protracted course (e.g., diabetes, hypertension). Chronic diseases are a major threat to human health and have complex causes. Such diseases lack specific evidence of an infection source, and some are not yet confirmed as diseases.

In 2012, Y Combinator, ranked in the Top 1 by *Forbes* among "Top Startup Incubators and Accelerators", invested in Healthy Labs,[24] a start-up devoted to serving patients suffering from chronic diseases. Healthy Labs' co-founders Sean Ahrens and Will Cole both studied computer science at the University of California at Berkeley.

---

[22]http://www.wma.net/en/30publications/10policies/b3.
[23]http://grants.nih.gov/grants/guide/rfa-files/RFA-FD-14-089.html.
[24]http://techcrunch.com/2012/08/20/yc-startup-healthy-labs-wants-to-be-the-go-to-site-for-people-living-with-chronic-illness.

Healthy Labs made its debut with the release of Crohnology – a social network for people dealing with Crohn's disease and colitis – in order to collect, share, and compare symptoms and therapies. To avoid medical service advertisements, users can only join after proving that they are suffering from this disease. And for privacy considerations, no data on the website can be retrieved by search engines. Healthy Labs aims to become a mediator in facilitating clinical trials of new drugs.

Monitoring in chronic disease includes "check on the progress or regress of the disease and the development of complications."[25] In this regard, data innovations include, but are not limited to the following:

1. *Exploring the Evolutionary Path of Chronic Disease.* In mining historical diagnostic records of chronic diseases, it may be possible to observe the evolutionary path over time and to design more scientific based treatments for patients' first visits, examinations, education, diets, medications, therapies, and nursing care to lower the incidences of blindness, randomness, and also repetition of unnecessary physician and healthcare services, so as to make the best use of the limited medical resources, enhance the quality of medical services, and reduce medical expenses.

2. *Chronic Disease Evolutionary Data Mining.* By analyses of data derived at different disease stages, there may emerge evidence as to the origin of the chronic disease and, by its progression, information as to how it can intrude the digestive, immune, respiratory, nervous, circulatory, endocrine, urogenital, and skeletal systems, thus providing knowledge as to the treatment of the chronic disease. At same time, this may help our understanding the interactions among different chronic diseases, and also the early warning signs of a chronic disease.

3. *Mining of Rare Disease.* Through abnormal group mining of electronic medical records, it may be possible to find rare diseases (e.g., Prader-Willi Syndrome) that have similar symptoms with other chronic diseases, which may only be observed in small number of people, or in those patients indicating abnormal curative effects under the same course of therapy, in order to optimize the original therapeutic method. Such findings could further be used for a curative effects analysis of the chronic disease.

4. *Correlation between Combination Therapies.* Analyses of connections between medication data and patients' medical health records may help medical institutions plan appropriate medication regimes and, subsequent to finding interactions among drugs, to explore variances in therapeutic drug medications and the incompatibilities among certain prescribed drugs.

5. *Medicare Data of Chronic Diseases Detection.* From data on Medicare payments, first, from the hospital's perspective, it may be possible to learn the physicians' prescribing habits in treatments of chronic diseases, and thus to ensure the reasonable use of Medicare funds. Second, from a regional perspective, it may be possible to study the geographic distribution of a certain chronic

---

[25] http://www.ncbi.nlm.nih.gov/pmc/articles/PMC554914.

disease, so as to inform susceptible populations on prevention measures while assisting patients who must receive uninterrupted treatment.

5.5.4   Data Supporting Data: Brain Sciences and Traditional Chinese Medicine

Regularity and high fidelity are two distinct characteristics of the scientific method, both of which must be built on observable, empirical, and measurable evidence. Yet, traditional research methods (e.g., theoretical research, normative research, empirical research) cannot prove with scientifically that the human brain consists of billions neurons and trillions connections, or that the traditional Chinese medicine (TCM) is based on mystical "meridian flowing" causes.

In January 2013, the European Commission announced an award for the flagship program "Future and Emerging Technologies" (FET)[26] to the "Blue Brain Project" led by Henry Markram at the École Polytechnique Fédérale de Lausanne. The project will receive scientific research funding of 1 billion euros over the next ten years. Markram will be responsible for coordinating 87 institutions or organizations, among them the Institut Pasteur (France), IBM (US), and SAP (Germany), in attempting to simulate the human brain and all its brain cell activities in order to study the chemical characteristics of and functional connections within brain regions. In April 2013, US President Barack Obama announced,[27] at the White House, the allocation of US$100 million for fiscal year 2014 to the Brain Research through Advancing Innovative Neurotechnologies (BRAIN), which includes participants from the National Institutes of Health, the Defense Advanced Research Projects Agency, and the National Science Foundation.

It should be pointed out that both research methods are based on brain data integration, analysis, and verification [5]. More specifically, the first step is to instantly or periodically acquire brain data through brain imaging techniques (e.g., fMRI, OT, ERP/EEG, MEG); second, to carry out longitudinal or lateral integration, with longitudinal integration indicating to integrate different levels of data for the same question, and lateral integration indicating to integrate the same level of data from different studies; third, to design new mining algorithms that can analyze the brain data in order to extract the hidden laws of brain cell activities in human problem solving, reasoning, decision making, and learning; and fourth, to verify whether correct knowledge about brain cell activities has been obtained.

In the field of traditional Chinese medicine, people use acupuncture needle stress monitors to record the twisting forces and sticking forces in the process of acupuncture therapy, and then to study the meridian through data analysis. By association analysis, Chinese medicine attempts to find evidences that the "five sense organs correspond to five internal organs," which is to say, the kidneys correspond with the ears, the heart with the tongue, the liver with the eyes, the lungs with the nose, and the spleen with the mouth (lips) [50]. Some practitioners even try to connect data on the properties, origin, and preparation method of traditional Chinese medicinal

---

[26]http://europa.eu/rapid/press-release_IP-13-54_en.htm.
[27]http://www.livescience.com/28354-obama-announces-brain-mapping-project.html.

materials with the data on herbal prescriptions, and thus design new anomaly detection or abnormal group mining algorithms to find the incompatibility of drugs in a prescription (e.g., the eighteen incompatible medicaments). These are all perceived as valuable research methods.

Therefore such data innovations that are "using data to support data" do get sufficient scientific evidence that contributes to advances in Chinese medical practices.

5.6   TRANSPORTATION DATA SERVICES

Data innovations are also being introduced in the transportation field. Unlike the overt commercial value in the healthcare industry, data mining in transportation visibly documents government initiative in performing public services. This is because transportation is generally critical to people's livelihood, in that transportation services are integral to the economy, society, culture, and politics.

5.6.1   Household Travel Characteristics

Household travel activities can provide valuable information for the government in developing transportation strategies. In the past, social surveys were used. The earliest survey can be traced back to 1944, in the final phase of the Second World War, when the US government recognized the importance of highway construction for national defense, and proclaimed the Federal-Aid Highway Act. Given the lack of household travel information then that could support highway planning, a survey method of "Home-Interview Origin-Destination" was proposed. In this survey, "households were asked about the number of trips, purpose, mode choice, origin, and destination of the trips conducted on a daily basis."[28] It strictly defined that "one trip" as a one-way trip of more than 400 meters or walking time of more than 5 minutes, so as to exclude inner streets, yards, or campuses that might have impact on the flow of traffic on urban arterial roads.

Today, data on transportation can be launched using specially designed multi-angle fast algorithms for mining data across multiple population clusters, vehicle ownership, land configurations, environmental changes, weather conditions, and traffic patterns, so as to obtain a picture of household travel activities without questionnaire surveys. For example, classification or cluster methods can subdivide the population based on trip purposes such as commuting for work or school, frequency of entertainment travel, or shopping at distant malls. Additionally, to identify the tidal flow of traffic, rule-based classifiers can be used to analyze the number of vehicles on roads at a specified time period, including flow of travel direction, distance, frequency, distribution, origin, and destination. Yet another consideration may be to utilize anomaly detection or abnormal group algorithms to analyze data on the lay of the land for road construction, environmental data for traffic pileups, and meteorological data for the seasonal upsets due to wind, rain, or snow. Based on such data, government can

---

[28]US Department of Commerce (1944) Manual of Procedures for Home Interview Traffic Studies. US Bureau of Public Roads. Washington, DC.

further understand (1) the demand of household travel arising from population expansion or economic development, (2) the levels of motorized traffic due to continued outward expansion of the city, and (3) the configuration of passenger revenue for rail-centered public transport. Moreover government could also do a reverse analysis, such as measure the layout of urban amenities according to household travel characteristics. Likely the results that would be used are still beyond our expectations as to what can be mined from these data.

5.6.2 Multivariate Analysis of Traffic Congestion

Traffic congestion has impacts ranging from reduced vehicle speed, increased exhaust emission, delayed time, or induced accidents to polluted environment; traffic pileups can weaken the magic of city attractions and even stall regional development. Of course, the extent of traffic congestion varies from city to city and from region to region. For example, the United States (Chicago) defines traffic congestion as "if the roadway occupancy is over 30% or waiting time is more than five minutes," whereas Japan defines fast lane congestion as "if vehicle speed is less than 40 km/h or vehicles' creeping and stopping exceeds 1 km and continues for more than fifteen minutes." Even so, waves of congestion worry government most during traffic reliability evaluations, next to natural disasters and facility malfunctions.

Abrupt or regular traffic congestion can result from conflicting transportation supply and demand. The many dynamics additionally involved are shown in Figure 5.1.

Only by understanding all factors contributing to traffic congestion can government take appropriate measures to curb its occurrence, instead of blindly singling out certain factors and not others.

In the past, indexes were used to measure associations between certain variables in analyzing probable congestion-contributing factors. For example, outliers were used to determine the extent of expressway congestion based on toll collection and also without toll collection. In other words, the attempt was to measure a connection between variables based on a minor data set that can only offer "verification" and not "falsification."

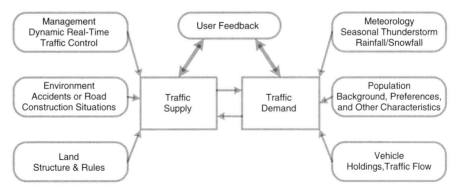

Figure 5.1   Possible congestion contributing factors

Application of data mining tools to multiple data resources can help reveal the correlations between every congestion-contributing factor and the supporting evidence for each. Such data resources should include population growth, vehicular, geographic, environmental, meteorological, and traffic management factors. For example, for infrastructure construction, government will be able to analyze historical data on construction sequence and timing of construction projects launched by different departments within the same road section, including power supply, water supply, communications, and urban construction, so as to maintain unified planning. Subsequently, for the urban area traffic, government will be able to analyze specific distribution data and take corrective actions where there is non-motor vehicle parking confusion, disorderly street parking, double parking, or public transport bus routes with irregular stops, and make readjustments such improving the distribution of bus stops, taxicab stands, and vehicle parking spots for both non-motor and motor vehicles. For household travel, government will be able to initiate re-planning after considering data on the most congested intersections, especially "which traffic signs or signals are improper arrangement or incorrect installation," and merge this with data on vehicle and pedestrian flow in different time frames.

5.6.3 Short-Term Travel Time Estimation

To be sure, a transportation system is a complicated system. To mitigate traffic congestion and environmental pollution, "intelligent transportation" can be instituted and travel time forecasts made to implement dynamic control and real-time management.

Travel time can be studied from number of perspectives: (1) the road network, whereby travel origins and destinations and passed entrance ramps are considered as "vertices" and passed road sections as "edges" in a road network, and then travel time analyzed using graph theory; (2) the traffic flow, whereby moving vehicles on the road are considered as "traffic flow" and "interrupted flow," when there is pause from a change in road shape and the control actions of signal lamps (an uninterrupted flow can be found on the expressway and the urban expressway), to analyze the impacts of (non-)linear routes or special points (e.g., the entry or exist of ramps) on the travel time; (3) the ambient conditions, whereby pre-travel time forecasting based on travel route planning is correlated with current traffic conditions using remote sensing, geographic, and fixed-position data, en-route travel time estimation; (4) the weather reliability, whereby added to the analysis of all road network features and traffic flow are seasonable issues.

No doubt, there will be in regularities due to cycle or successive differences when the time range is a week, a month, one season, or even a year. A longer forecast period can, technically, better results, as is true for all hypothesis-based models, as would be the case of predicting the future traffic situation by building mathematical or physical models fitting a mechanism formed from certain traffic phenomena in order to indirectly obtain travel time. In a short period (e.g., the next one hour), for which there will be strong uncertainties, travel time estimation is potentially more challenging but useful for travelers in (1) selecting a transportation mode, (2) making route choice decisions, and (3) determining departure time from a place.

Early short-term travel time estimation was based on models that generally could not deal with the impacts from random interference factors. These include the Kalman filter, nonparametric regression, Markov chains, and artificial neural networks. The Markov chain model, for example, describes the probability of delay in travel time through a transition matrix, but the parameters are obtained under limited test conditions and may differ significantly from actual conditions. However, due to the complex nature of the transportation system, fitting based on a "perfect" model is not feasible.

However, data-based studies on short-term travel time estimation can be (1) simple and effective, since mining algorithms can simultaneously evaluate thousands of hypotheses and boundary condition settings using the multilevel adaptive iterative method, and are more convenient than conventional single-model-based methods; (2) easy to handle, since multi-source data sets related to transportation data involve population, vehicle, land, environment, meteorology, and traffic management, and additionally are heterogeneous, complex, high-dimensional, and distributed, all of which cannot be handled by traditional methods.

5.7 TRANSACTION DATA SERVICES

In recent years, the innovative feature of commerce has been to extract partial value-added activities to shape transaction services relating to pricing, sales, and payment.

5.7.1 Pricing Reform

Traditional pricing of commodities generally concerns how to (1) target profit pricing, (2) estimate price elasticity of demand, (3) estimate the cost, (4) carry out competitor analysis, (5) select a pricing method, and (6) calculate the final sale price. Targeting profit pricing must give consideration to consumer behavior, which can be related to the consumer's class and social or cultural background. Estimating price elasticity of demand must give consideration to bottom price, ceiling price, cost-efficient price, and high-value price in order to maintain a presence, maximize the current interest, maximize market share, and maximize product quality. Cost estimation must give consideration to the overall cost consisting of fixed cost, variable cost, marginal cost, and opportunity cost, while competitor analysis intends to study the competitors and their products on the market, with more competitors bringing down the price. Selecting a pricing method is to know the competition, demand, and cost adjusted pricing methods. The calculation of final price must take into account the discount pricing strategy (e.g., cash, quantity, functional, seasonal, price discounts), regional pricing strategy (e.g., discount for a specific region), psychological pricing strategy (e.g., prestige, mantissa, loss leader pricing), new product pricing strategy (e.g., skimming, portfolio, penetration pricing), and discriminating pricing strategy (e.g., pricing based on customer segmentation, product positioning, location or time). But, overall, pricing is based on consumer behavior, and certainly there are strategies that can confuse consumers. These strategies, such as the contrast effect (taking on significance from neighboring prices) and the anchoring effect (a particular number such as "nine"), had been summarized in William Poundstone's best-selling book *Priceless: The Myth of Fair Value* [51].

Online retailers, although not having to pay lease and business rates, which gives them a cost advantage, nevertheless face drastic price competition the same as the traditional trade. This is because online retailers need to apply additional variable inventory control and variable pricing. Variable inventory control refers to replenishing inventory based on marketing strategies or merchandise sales due to limited space. Variable pricing means to proactively reduce or increase the price according to consumer demand, competitor response, or producer output.

The near future will bring a radical pricing reform based on data innovation that will reverse these practices, namely merchandise prices will be set by consumers instead of by retailers. As early as a decade ago, the website PriceLine.com had tried this method, and hailed an unprecedented success in the stock market – its shares gained 80% from the initial US$16 per share IPO price, and its total market capitalization in April 1999 reaches US$11 billion. PriceLine's business is to collect and release prices of air tickets, hotel rooms, automobiles, and house mortgages acceptable to consumers on its website, and then wait for the most appropriate vendors. Even now, transaction data mining is getting more convenient than earlier approaches in meeting the needs of all parties and market dynamics. Just think: it is beneficial for both the airline and consumer when an airfare flight to Europe is offered to a consumer who happens to want a trip to a similar destination at a desirable price.

5.7.2   Sales Transformation

Invented by Jeff Bezos in 1994 and redefined by IBM in 1995, e-commerce has been developing for about twenty years as an advanced sales mode in the commercial distribution area. However, this is just a simple reflection of traditional trade digitization. To be a real sales transformation, e-commerce has to continue to change – a change that depends more on the C2B (customer to business) model. Other models as B2B, B2C, and even B2G (business to government) exist in traditional trade and rely on the "large-scale, assembly-line, standardization, and low-cost" standards in the traditional industrial age; especially, the C2C, despite its huge impact, still has offline traces. Meanwhile inventory is a common issue critical to all the above three models.

Chris Anderson argued, in explaining his long-tail theory, that "products in low demand or that have a low sales volume can collectively make up a market share that rivals or exceeds the relatively few current bestsellers and blockbusters, if the store or distribution channel is large enough" [52], and even larger. In the future commercial distribution, the value chain will be driven by consumers instead of manufacturing or design. This is the essence of a real sales transformation. As a matter of fact, the C2B model does not reject generality, and can be roughly divided, on the one hand, into demand aggregation, such as reverse group buying, and on the other hand, into offer-based cooperation, such as reverse auction. Technically speaking, these are both innovations based on transaction data, and so require more efficient partition, analysis, and locking of consumers. In other words, the C2B model allows follow-up improving commodities and promoting sales according to the niches' micro demands, which is indeed a bespoke transform development from "large scale" to "personalized, multiple-varieties, small quantities, and rapid response."

To be sure, the world's economy since the global financial crisis in 2008 has faced a long and arduous road to recovery, and there was an urgent call for transformation and upgrading. Apart from gaining new competitive advantages through "technology, brand, quality, service, and efficiency," it has become clear that commerce can no longer promote B2C with homogenized products. Instead, retailers must focus on the individual demands of consumers as in the C2B model, in order to drive flexible and efficient industrial methods of manufacture.

5.7.3 Payment Upgrading

As the cyber-consumption grows to a size that was never seen before, so has the competition of non-cash in the global payments landscape, as evidenced by "the significant amount of venture capital – roughly US$76 billion – that has gone into payments-related business since 2010" according to the Global Payments 2015[29] released by Capgemini and the Royal Bank of Scotland.

True payment upgrading will come with both sizable disruption and immense opportunity, and so be buttressed by data innovation.

1. *Expanding the use of the virtual currency with payment methods.* There are now more than 70 virtual currencies distributed worldwide, with the largest players being bitcoin, ripples, and litecoin. Regardless of what your concept of "real" money is, the fact is that these virtual currencies are already operating as legal tender. As earlier in 2014, the US Internal Revenue Service (IRS) issued a notice that declared "the virtual currency is treated as property for the US federal tax purposes."[30] Meanwhile, mining of the virtual currency is becoming an increasingly institutionalized pursuit driven by multimillion-dollar businesses. This trend will require newer technologies or faster algorithms to significantly shorten the processing time; for instance, Coinalytics that announced having raised US$1.1 million as part of a seed round on September 2015, helps clients to conduct risk assessment in the bitcoin space, including "unnamed payment processors, wallet providers and exchanges."[31] Another interesting start-up Coinometrics is attempting to "empirically quantify and qualify the actual behavior happening on the blockchain,"[32] which is not just a focus on the currency itself.

2. *Transforming currency-based payment into credit-based.* The transformation will further promote the development of national and regional credit systems, covering both individuals and enterprises. For example, Alibaba Financial launched a "virtual" credit card for clients who have good credit

[29]https://www.worldpaymentsreport.com.
[30]https://www.irs.gov/uac/Newsroom/IRS-Virtual-Currency-Guidance.
[31]http://letstalkpayments.com/blockchain-technology-based-company-coinalytics-raises-1-1m.
[32]http://www.businessinsider.com/10-most-promising-blockchain-companies-2014-5.

ratings computed by their semi-/or automatic classification system, with a maximum application quota of 20,000 RMB and an interest-free period up to 38 days for purchasing at Tmall.com. The corresponding money comes from a short-term capital pool formed by other Alipay users, and the profit is made though the charge of handling fees.

6

DATA SERVICES IN DISTINCT SECTORS

Although the terms "sector" and "industry" are often used interchangeably to describe a group of enterprises that "operate in the same segment of the economy or share a similar business type," they do actually have slightly different meanings. Here, we also adopt the terminology used in the stock market in referring to a sector as the broader classification. In addition, the terms "domain" and "field," used in Chapter 6, are most often used to designate an area of professional specialization.

Currently, enterprises or organizations that own or control data resources do not participate in data innovation. In fact, because of the usual barriers to entry, it may not be possible or realistic for them in the sectors to join in making data innovations. As Kevin Kelly wrote in his book *Out of Control: The New Biology of Machines, Social Systems, and the Economic World*, we might apply "bee thinking" as well in referring to the data industry, and crowdsource projects using a "mini-plant" groups consisting of 8 to 12 persons, so as to realize such a "distributed, decentralized, collaborative, and adaptive" "co-evolution." Transboundary (or trans-sectoral) cooperation, therefore, can be consider as an important means of discussing possible future direction for data services like the issues of fostering SMEs and entrepreneurship development. We know all the scenarios in this chapter will prove beneficial to transboundary cooperation.

6.1 NATURAL RESOURCE SECTORS

The Earth's surface is a complicated, secondary, huge system. It is primary where in a distribution of materials, energy transformation and material transfers occur, involving the atmosphere, lithosphere, hydrosphere, and biosphere, which do not provide habitable space for the survival and reproduction of species but do provide many natural resources for the life, production, and social development of humankind, including the circulation of water, land, mineral, and energy resources. Natural resource sectors have the capabilities of using the Earth's resources to generate revenue and create jobs.

6.1.1 Agriculture: Rely on What?

From time immemorial, agriculture has been the fundamental sector guaranteeing a people's livelihood to avoid being hungry. Everyone thinks it does exist and will remain irreplaceable until any real substitute stuff can be found through a highly developed science and technology. Agricultural products in the futures markets, the same as energy, metals, and their derivatives, also have an asset value, these including soybean, soybean oil, bean pulp, paddy rice, wheat, corn, cotton, white sugar, cocoa, coffee, tea, rapeseed oil, palm oil, banana, and rubber.

Because in agricultural production humans interact with nature to obtain resources, throughout the ages agriculture has been regarded as a profession that "relies on the weather." The ancient Egyptian farmers divided their year into three seasons, based on the cycles of the Nile River, as early as ten thousand years ago, and the "farming season," as well as its significance, was realized early even in ancient China. Over the last several decades, as technology has rapidly progressed, there has developed a consensus that agriculture can bid farewell to seasonal issues. Under the aegis of meteorological science, traditional dependence on the weather has evolved into observing the weather. However, agriculture as served by meteorological science has come upon two disadvantages. First, low popularization, which is due to agro-meteorological services being usually applied to scientific research, and there is still the "last mile" gap before any practical application can emerge. Second, much too short forecasting, which is due to forecasting doing relatively better in the 6 hours to 15 days range. In the recent years, driven by the increasing investment in the Internet of Things, through data capture of wireless sensor networks, agricultural applications have started to close the "last mile" gap in meteorological services. An optimistic finding is that data capture and monitoring will be changing from observing "the weather" to observing "data."

Yet, it is not enough to only observe the data by manual monitoring. We need to mine and analyze data. The Climate Corporation,[1] founded as WeatherBill in 2006 by two former Google employees – David Friedberg and Siraj Khaliq – can ingest weather measurements from 2.5 million locations, combed with crop root structures and soil analysis, to simulate the climate and locate extreme and unusual weather events. In June 2012, Climate announced a US$50 million dollars of Series C funding with investors Khosla, Google Ventures, and the Founders Fund. The answer here is quite plain: let agriculture rely on the data as well as technology.

---

[1] http://techcrunch.com/2012/06/14/founders-fund-leads-the-climate-corporations-colossal-50m-funding-round

6.1.2   Forestry Sector: Grain for Green at All Costs?

Aldo Leopold, an American environmentalist, stated in the concluding essay in *A Sand County Almanac* that the more slightly the man tampers with nature, the more effectively the "land pyramid" will modulate adaptation. Essentially, the grain for green problem is not slight tampering but a violent transformation.

Leopold's experiment was used by Kevin Kelly as a case[2] of restoration biology. In 1934 Leopold purchased a worn-out and abandoned farm from the University of Wisconsin and started to restore it to prairie. However much he plowed and sowed, the prairie was still overgrown with grass and weeds. After ten years he eventually knew that "fire" was the key to reorder the whole community ecology. This case's sweet hereafter is that in 1983, Steve Packard reproduced an ecosystem similar to prairie – "sowed the mounds of mushy oddball savanna species" at random, but "within two years the fields were ablaze with rare and forgotten wildflowers: bottlebrush grass, blue-stem goldenrod, starry champion, and big-leafed aster." In 1996, Stuart Pimm and Jim Drake, ecologists at the University of Tennessee, revealed the importance of evolution sequence with assembly experiments and validated Packard's work in restoring the savanna. Packard's first try failed in the sense that "he couldn't get the species he wanted to stick and he had a lot of trouble taking out things he didn't want." However, once he introduced the oddball species, "it was close enough to the persistent state." The savanna was accidental but in fact inevitable based on the assembly sequence.

Grain for green does not just depend on our will. In practice, prior to carrying out the work with human financial and material resources, data analysis is needed. One is to calculate the costs and benefits of all interested parties as it's unnecessary to do this for fertile regions. The other is to construct with algorithms a simulation of a simple ecosystem of the region that needs to be returned, and to select the appropriate way to return the farmland.

6.1.3   Livestock and Poultry Sector: Making Early Warning to Be More Effective

Livestock and poultry is a sector that requires the most epidemic prevention. Epidemics are a constant threat concerning three types of outbreaks. First are the pest outbreaks, which vary in distribution, transmission route, quantity, and hazard status, and destroy seedlings and crops. Second are the farm animal disease outbreaks, of which the common diseases are Salmonella gallinarum, Mycoplasma gallisepricum, and avian leukosis viruses, and thus spread as major zoonotic outbreaks of bovine spongiform encephalopathy (or mad cow disease), foot-and-mouth disease, and Avian influenza (or bird flu). Third are the fish and wildlife disease outbreaks, which by their migrating routines may acquire and host pathogens that can be transmitted to susceptible livestock. An early warning system is necessary to prevent an explosive outbreak at the source and ensure a rapid, scientific, orderly, safe, and effective response.

---

[2]http://kk.org/mt-files/outofcontrol/ch4-b.html

Currently, the early warning system is implemented by combining an epidemiological survey, sampling inspections, and a clinical diagnosis. For instance, immunity sampling, antibody detection, and etiological tracking of an individual animal, collection and processing of major monitoring objects from the survey data, and releasing forecasts after a quasi-real-time tending-to-macro evaluation.

The accumulated data on the epidemic and the outbreak warnings are then adjusted from simple real-time data monitoring and collection to data mining in combination with all available data resources. A recent example is the H7N9 influenza. Although Google had not involved Chinese data in its forecasting scope, the Google influenza warning service still predicted that the flu outbreak period in Russia, United States, and Canada in the Northern Hemisphere was to be late winter in December 2012 to early spring in March 2013, and was to be from June to August 2013 in the South Hemisphere. The two periods highly overlapped with the migration of migrant birds. Among the eight main pathways, five went through Russia, and one of the five started from Alaska over the West Pacific Islands and near the eastern coastal provinces in China. The major epidemic region of this H7N9 outbreak was in the eastern coastal provinces of Zhejiang, Shanghai, and Jiangsu.

6.1.4   Marine Sector: How to Support the Ocean Economy?

According to a recent forward-looking assessment of the OECD (2012),[3] the ocean economy has huge development potential. The marine sector could be developed to include "off-shore wind, tidal and wave energy, oil and gas extraction in deep-sea and other extreme locations; marine aquaculture; marine biotechnology; sea-bed mining for metals and minerals; ocean-related tourism and leisure activities; and ocean monitoring, control and surveillance." All the while, the "traditional maritime and coastal tourism, and port facilities and handling" course of marine life would not be disrupted. However, exploring the oceans as well as deepwater development produces enormous technological challenges and poses serious threats to the environment. Thus the question keeps coming back to the timeworn economic issue of how to balance economic development and natural resource use.

Today, we have massive oceanographic data in the marine science field, including submarine relief data, sea level data, ocean remote sensing data, ocean data assimilation and modeling, boat test data, buoy data, tide station data, and various oceanographic survey data. These data resources are widely shared by the scientific community. In the main, these include 2-Minute Gridded Global Relief Data (ETOPO2v2) of the US National Oceanic and Atmospheric Administration, the International Comprehensive Ocean Atmosphere Dataset (ICOADS) collected by the US National Climatic Data Center, the coupled ocean-atmosphere mode data set provided by the European Centre for Medium-Range Weather Forecasts, the oceanographic survey data set acquired from the Array for Real-Time Geostrophic Oceanography (ARGO), and the Tropical Ocean Global Atmosphere-Coupled Ocean Atmosphere Response Experiment (TOGA-COARE).

It should be noted that without the development and utilization of oceanographic data, there can be no transformation of the ocean economy to quality economy, marine

[3]http://www.oecd.org/futures/Future%20of%20the%20Ocean%20Economy%20Project%20Proposal .pdf.

exploitation to cyclic utilization, marine science to leading innovation, and the marine rights to overall consideration. Oceanographic data innovation is the only way to create an economic strategy for exploring the oceans, and this would entail high technical support for the development of the ocean economy, the protection of marine ecology, and the maintenance of maritime rights and interests.

6.1.5 Extraction Sector: A New Exploration Strategy

Despite the Club of Rome's sounding its continual alarm in *The Limits to Growth* and noting the paradox of the resource curse, it cannot be denied that extraction sector is basic to the globe economy in meeting present and future needs. But we need to face up to the inevitable: some natural resources will be depleted all the more quickly because of a current consensus to efficiently maximize resource exploitation in accord with the utilization needs and via new technologies to minimize the ecological damage.

According to the *Oil and Gas Industry Collaboration Survey 2011* by Avanade,[4] a joint venture company of Accenture and Microsoft, to acquire more fortune and value, many oil and gas enterprises should apply data mining technologies to the tasks of strategic decision making, science and technology R&D, production and management, and safety and environmental protections. BP Global owns a powerful world-scale data center (3-story and 110 K sq ft) that can provide standard management and outsourcing services. Royal Dutch Shell has established a petroleum data bank system for management of its data assets based on OpenWorks grids. The Chevron Corporation has introduced data sharing at 50,000 desktops and 1,800 company locations to eliminate repeated processes in "downstream" systems of oil refining, sales, and transportation, and has had an accumulative return of US$2 billion net present value. In addition Schlumberger, Halliburton, and Baker Hughes, by building a research and working team for data exploration and development integration, have supported the oil field production planning and decision making, in order to enhance oil and gas resources development in unconventional areas such as deep water and polar regions.

Such a big data actions by these oil and gas giants, in essence, signals the value of new data exploration strategies in developing high-grade or hidden resources. In geophysical and geochemical exploration, there are three categories of geophysical data. The first category is petrophysical data; this data comes from the research of rock frames, dynamic processes in the earth, mantle convection, the geological environment and its evolution, and includes permeability, saturation, electricity, mercury penetration, clay mineral, casting section, particle size distribution, which is a direct or indirect connection with aquifers. Second is logging data; this data can be used for correct identification of fluids, including porosity, permeability, water saturation, oil saturation, and water production rate, which is mainly collected or extracted by traditional exploration means. Third is geologic data like seismic frequency; this data can be used to judge interlayer sliding and tectonic faulting, to direct mineral resource exploration using specific depositional environment and structural features.

---

[4]http://www.accenture.com/SiteCollectionDocuments/PDF/Accenture-Upstream-Software-Solutions-2011-Summary.pdf#zoom=50.

Obviously, data mining tools in the extraction sector is no longer aimed merely at geological interpretation. The data provide insight as to the interior and unrecognized connections among the various, vast geological data so that new methods of "direct" resource exploration transformed from "indirect" supplementary means can be applied to enhance growth of this sector.

6.2   MANUFACTURING SECTOR

Industrialization, needless to say, is a source of tremendous progress in materials of human civilization. Industrialization is characterized by repetitive processing and the manufacturing of large quantities of standardized products in production lines, including but not limited to, iron and steel, metallurgy, cement, glass, paper, alcohol, leather, printing and dyeing, chemical fiber, textile, clothing, shoemaking, bags and suitcases, toys, electromechanical, furniture, and plastic products. Industrialization represents four types of developments: knowledge-driven, merchandise exchange, market expansion, and accumulation of capital.

Because of the environmental pollution caused by many of the industries in the manufacturing sector, a new industrialization path is being explored. The "new path" is a concept manifest, on the one hand, in abandoning the traditional view of developing, first, the economy and then, second, protecting the environment and, on the other hand, venturing out on a new technological path. It should be noted that data innovation meets both requirements.

6.2.1   Production Capacity Optimization

Production capacity is volume that can be expected in a given period under existing organizational technological conditions. Capacity is a technical parameter that measures manufacturing capability, and can also reflect the scale of production. Excess capacity means that "the demand in the market for the product is below what the firm would potentially supply to the market."[5] We usually use the capacity utilization rate, a major macro and micro indicator, to measure whether or not production capacity is in excess. If the capacity utilization rate is low, then that specifically indicates that enterprises should try to increase production. For instance, the United States takes 78% to 83% as a normal range for capacity utilization (excess capacity below 75% and under capacity above 90%), and Japan's is 83% to 86%.

Production capacity optimization has multiple issues that depend on different perspectives and various factors. In fact the concept of excess capacity is not strict – moderate overcapacity may be a preferred choice under information asymmetry and low elasticity of production factors. Optimizing capacity therefore does not imply blindly underproducing, and certainly cannot be applied in mechanically executed econometric models.

---

[5]http://www.investopedia.com/terms/e/excesscapacity.asp.

Data innovation will play a distinctive role in production capacity optimization. In *Wall Street Journal* report in May 2013,[6] for example, Raytheon, a defense contractor, is said to have acquired knowledge from data accumulated from its own prior practices and those of other suppliers in order to optimize production capacity as well as product quality by eliminating defects and tracking hazards tracking. From this, Raytheon's practice is a referential experience. In targeting such a problem of developing China's new industrialization path – namely optimizing China's production capacity in certain "structural overcapacity" sectors as iron and steel, coal, automobile, as well as real estate – We should break the limit of econometric models, and apply cross-sector data mining, to conduct top-level design, so as to arrange specific production capacity optimizations started with environmental requirements like carbon emissions.

6.2.2   Transforming the Production Process

Working capital is as essential for enterprises as "blood is necessary for life," especially for manufacturing enterprises. Generally, there are five types of costs in a manufacturing enterprise's routine operations: raw materials, production line, sales, inventory, and liquid capital.

The concept of "working capital" helps us make sense of certain suppliers, such as those of automobile parts and accessories, whose manufacturers' facilities are always located close to vehicle manufacturing enterprises. The SMEs, in particular, are biased in favor of participating in alliances of every kind to share business information, and with companies that operate under a pay-on-delivery model in fast-moving products. Compared with setting up relationships among distant suppliers, competitors, copartners, clients, and the others in the market sector, having all facilities nearby, and adjusting production accordingly, is a more efficient way for enterprises to lower the cost of working capital, as well as to resolve issues of resource dispersion, product backlog, excess inventory, and underutilized equipment.

Data innovations can transform the production process and also the overall structure of the workflow data. To be sure, the industrial community as a whole benefits from the accumulation of the workflow technology when data produced by task assignment is flowed or transferred among different activities, different executors, or different activity levels. In this capacity, workflow data can be classified as control data, relevant data, and application data, as would pertain to the original purpose of workflow data mining, which is to "usefully abstract the actual process executions and capture important properties of the process behavior" [53]. Workflow data mining can deduce business context, hidden patterns, and rules from frequently executed pathways, infer correlations between business activities from the minimum support thresholds of critical triggered factors, and so forth. For instance, data mining may find a more practical bottom line price (e.g., 100 dollars) for some materials in the material procurement process, or a way to control the amount of raw material needed to produce certain products in the manufacturing flow, and may even

[6]http://www.wsj.com/articles/SB10001424127887324059704578472671425572966.

effect a warning to drop production in order to prevent product backlog or excess inventory.

6.3 LOGISTICS AND WAREHOUSING SECTOR

Logistics, as defined by Council of Logistics Management (1991), is a "process that plans, implements, and controls the efficient, effective forward and reverse flow and storage of goods, services, and related information between the point of origin and the point of consumption in order to meet customers' requirements." The general description of warehousing is "performance of administrative and physical functions associated with storage of goods and materials" (Businessdictionary, 2007). The key functions of warehousing operations[7] are receiving, conveying into storage, order picking, and shipping.

Logistics and warehousing are nearly always closely connected – both as value-adding service providers and terminal business customers. Many of the world-famous industrial warehousing operations, including Singapore-based Global Logistics Properties, US-based Gazeley, Australia-based Goodman Group, and Japan-based NewCity Corp, are engaged in logistics. Additionally there are such specialized logistics enterprises as Chemion, Kruse, and Talke, which are good in the storage of chemical products.

6.3.1 Optimizing Order Picking

Among the warehousing functions, order picking is quite important because of the complicated procedure of meeting customer orders. It has been estimated [54] that order picking accounts for up to 50% of the total warehouse operating costs. Order picking consists in "taking and collecting articles in a specified quantity before shipment to satisfy customers' orders."[8] However, more precisely, identifying the pros and cons of order picking is whether it takes a long operational time, which is composed of order preparation time, picker travel time, goods searching time, and goods picking time.

In the past, due to low amount of information used in warehousing, order picking consisted largely of manual operation subsuming pickers' proficiency and travel time. In fact, as early as 1959, American mathematical scientist George Dantzig proposed the vehicle routing problem (VRP) [55] to determine the optimal picking route. Since the VPR could substitute in the harshest criteria for the assessment indicators of the pickers, this problem drew the attention of many scholars as to contriving a routing traversal strategy with the shortest routes. Overall the methods were quite identical but with some interesting minor differences. First was to generate a hypothesis, including warehouse space structure, placement of goods, and order requirements. Second was to select a model embedding different algorithms, such as heuristics, an ant colony, a particle swarm, a neural network, and a genetic algorithm. Although

[7]http://www.cl.cam.ac.uk/~mh693/files/undergraduate_thesis.pdf.
[8]http://en.wikipedia.org/wiki/Order_picking.

some of the optimization experiments showed good results, they were partial toward theory and not so good for practical applications.

Today, we have data mining technology for applications that can help warehousing enterprises optimize the entire process of order picking. In the picking process, for instance, data mining can find the hidden patterns, and infer rules on coordinating operations. In the warehouse layout, data mining can dig for signs of customer preferences based on their goods selection to predict future orders. Last, in the placement of goods, data mining can calculate the shelf lives of goods, determine the piling height, of goods by learning stable positions them placed on the real warehouse environment and estimate access frequency based on correlations among different goods.

6.3.2 Dynamic Equilibrium Logistic Channels

For shippers the critical factor may be to choose a logistic enterprise with dynamic equilibrium channels. Non-equilibrium logistics channels or frangible connections can, during traffic congestions, raise costs of transport due to low-speed oil consumption, environmental pollution, time delay, or traffic accidents, and thus distort as well as social costs in resource allocations with subsequent social welfare losses. Research shows that congestion costs can be huge. Take the total social costs associated with the port of New York and New Jersey expansion as an example, and the increased hinterland highway traffic volume: each new increase of 6% container traffic can cause annual congestion costs to increase by 663 million to 1.62 billion US dollars.

Dynamic equilibrium channels belong to the spatial analysis of logistics, and thus provide a strong foundation for logistics planning. Most of existing research is based on location theory for enterprises. Usually, this is conducted by supposing a core location point in space and selecting a static model, by the notions of location timeliness, coverage, and compactness, and in combination with sections and transport lines selection. It is an empirical analysis of theoretical economics.

Due to a barrier to entry into the logistics sector, there were very few studies purely on logistics data mining that focuses on equilibrium channels. This is because, on the one hand, it is difficult to collect the data and, on the other hand, there is a small amount of data to be collected. It should be noted, however, that this area has great prospects, as reflected in the following two situations. One is in emergency response logistics in the case of a huge environmental disaster, calling for disaster recovery logistics to meet the demand for services. The other is to promote corporate transformation and upgrading, from providing simple segmented services to offering a complete selection of flexible services and price transparency.

Returning to the warehousing case, in merging the logistics and warehousing data sources, graph mining should be used. The warehouses would be extracted as vertices and the logistic channels as edges. Then the multiple relations would be correlated between the two sectors to study their synergy and to optimize the solution.

6.4 SHIPPING SECTOR

Shipping traditionally signified a nation's far-reaching economic prowess. Today, there are three dominant modes of transport:[9] aviation, maritime transport, and land

---

[9]http://en.wikipedia.org/wiki/Mode_of_transport.

transport (rail, road, and off-transport). Here, we consider only the new opportunities for shipping.

6.4.1   Extracting More Transportation Capacity

The construction of a critical modern international maritime port is not just an end in itself, but a way to encourage data innovations to solve practical problems at the operational level. A primary issue may be to increase the port's handling capacity as well as its operational efficiency. However, port handling capacity is not determined only by the number of deep water berths; capacity is also influenced by the number of port operations (e.g., port tally), the port deposit's storage tank conditions, the surface and air modes of transportation, and the logistics agency status (e.g., ship-owners, consignors as well as their agencies). Therefore, for the purpose of increasing port handling capacity, the main thrust here is to delve into transportation capacity based on existing port logistics data resources.

Data on transportation capacity may be extracted by the following two steps. One is to infer the rules of port development evolution, based on analyzing the changes of "port as well as its hardware configuration and software condition" using evolutionary data mining. The other is to find the correlations among supply and demand within shipping markets according to consumer preferences, so as to forecast market demand. Take the Port of Shanghai as an example. Currently, road transport has higher demand, but that does not mean we should reduce land transportation. The proper approach is, first, to merger the multiple available data, including the historical data on dredging the Yangtzer and Huangpujiang rivers, on building freight railroad, and on establishing cargo flights; second, to forecast by evolutionary data mining by collecting the distribution of speed at a given radius, with long and medium term, short term, and even immediate term predictions, so as to improve the elasticity of the port's handling capacity.

6.4.2   Determining the Optimal Transfer in Road, Rail, Air, and Water Transport

All the modes of transport – be they road, rail, air, or water transport – have advantages and disadvantages. Aviation offers high-speed, small traffic volume, and high transportation costs; maritime transport offers low-speed, large traffic volume, and low costs; rail offers large traffic volume but at low frequency, and trucking offers flexibility but is the most accident-prone. Therefore designing a good means of transport transfer of shipped goods requires careful planning.

Jules Verne, the "father of science fiction," described a very successful means of transport in his book *Around the World in Eighty Days*. In the story, English gentleman Phileas Fogg and his newly employed valet Passepartout are smart enough to travel by rail, steamer, balloon, and even elephant across the continents of Europe, Africa, Asia, and America to go around the world in eighty days. In real life, a similar event was attempted in 2013 during the Spring Festival travel season. Wang Dong,[10] a

---

[10]http://usa.chinadaily.com.cn/business/2013-02/04/content_16197439.htm.

native of Deyang, Sichuan province, and a PhD student at Fudan University in Shanghai, decided that rather than go through the hassle of buying a direct train ticket from Shanghai to Deyang. He would design a unique railway route home to avoid having, to buy train tickets during the holiday crush, by buying eight train tickets with five transfers. The direct train between Shanghai and Deyang takes 34 hours, whereas Wang's trip was to take less than 22 hours, traveling through Nanjing, Hankou, Dazhou, and Chendu before arriving at his final destination. Tickets on the direct trains from Shanghai to Deyang are in short supply. "I had barely any difficulty buying tickets in these sections," Wang said.

In earlier studies transport was usually treated as the interchange issue – a form of regional planning. This was because, from a temporal standpoint, changing transport mode adds to travel time. Thus, determining the optimal mode of travel transfer by road, rail, air, or water is easy to implement using data innovations that respond to the needs of time-sensitive or price-sensitive passengers. In addition data on transport transfers can be very helpful to the construction of critical international ports.

6.5 REAL ESTATE SECTOR

Although in 2008 there were witnessed many housing bubble bursts throughout the world, real estate continues to be a strong long-term investment and occasionally an income generator. This is because the real estate sector not only can guarantee national political stability but also stimulate economic growth. However, this does not mean that the real estate sector does not need change.

6.5.1 Urban Planning: Along the Timeline

Today, there is much talk that the whole world has entered an urban age, with "54 percent of the world's population"[11] residing in urban areas (that corresponding to rural areas) in 2014. Urbanization,[12] whereby humans migrate from rural areas to settle cities, is a societal phenomenon tied to a region's industrialization. For the emerging economies, urbanization raises issues of proper management of urban affairs and also urban design and planning. Urban planning is the more important issue for would-be residents, since in emerging economies appropriate conditions need to be in place to support enterprises and industries (which can boost the real estate sector as well), as historically been demonstrated by many advanced world cities.

Present-day urban planning procedures usually depend on spatial planning, which can be defined as "the coordination of practices and policies affecting spatial organization."[13] Spatial planning that involves qualitative content analysis, is relatively simple to breakdown of the urban space: its dimensional reach, topography, its ethnic groups, and existing trade routes. The idea is to first obtain a comprehensive picture of the locality and its hinterlands based on optimal points of interest. The next stage

[11] http://esa.un.org/unpd/wup/Highlights/WUP2014-Highlights.pdf.
[12] http://en.wikipedia.org/wiki/Urbanization.
[13] https://en.wikipedia.org/wiki/Spatial_planning.

of planning is to design a probable network of streets around natural "obstacles" and also leading to important resources. For example, there may "survival line" roadways planned to sustain farmlands at the city's extremities, and to access water resources; "ecological line" roadways that connect to ecological function zones and national nature reserves; "development line" roadways where areas are reserved for socioe-conomic growth in optimizing a people's rural–urban transitions; and a "guarantee line" roadway leading to the major energy or important mineral production bases and transport channels.

Because a spatial layout is conceived for construction over a future period of time, urban planning is subjected to a certain timeline. That is to say, different time periods of a region's development can be linked together, depending on issues of resource utilization or any matters relating to ecological restoration, relocations of a former town centers, and other municipal building projects. To aide in these instances there are some big data analytics like the Manhattan Software that can help the urban planners in some regions learn the histories of important residential properties and find ways to reduce costs or obtain immediate returns.

6.5.2 Commercial Layout: To Be Unique

A commercial property keeps the leasing (or selling) momentum going, depending on whether the physically the properties are well maintained. Yet, all the planning of architects is not enough. The most severe consequence of excessive growth is that shopping districts, whether core, secondary, or urban edge shopping districts have cookie-cutter layouts. The current practice in specialized commercial layout begins with a questionnaire that effectively helps architects understand such factors as population distribution, industrial structure, traffic situation, peer competition, and consumer preferences.

Commercial properties whether of flexible layout or unique architecture indu-bitably cannot go without people-centric data innovations. More sources of data may be needed to meet the multiple objectives that need to go into dramatically improv-ing the comprehensiveness, granularity, accuracy, and appropriateness of commercial buildings, to reach the broader range and deeper depth of analytical procedures. For example, the data innovation can explore the influence of shopping attractions or travel routes to a shopping area on a certain consumer group, or on family-style casual shopping, and thus design a novel the internal system of routes with several commer-cial points of interest.

6.5.3 Property Management: Become Intelligent

There is a prevailing view that property management is an extension of the real estate industry. To some extent this makes sense. At the initial point of market positioning a real estate project, property management has to have been established. In fact good property management can ensure the success of a real estate venture.

In the past, some unscrupulous developers would evade or shut down their man-agement services upon a building's completion or after all units are sold. The manage-ment that contracted to build a certain real estate thus pushed all the problems related to building maintenance, repair, and operations to unrelated enterprises. Such irresponsible behavior often resulted in disputes between the unit owners and property management enterprises, and with sufficiently large number of complaints even led to protest incidents. Fortunately, a few real estate enterprises have started to recognize the importance of "intelligent" property management.

What can intelligent property management do? The quality of a housing or commercial community is reflected in the property management services such as those of government. Examples are assisting households with energy savings (e.g., WaterSmart and Opower visualize water and electricity data to provide a complete household saving scheme), with security services (e.g., Nest corporation, founded by iPod designer Tony Fadell, offer smart thermostats to dynamically tune the indoor temperature and test for home invasion by temperature induction), or with remote care, (e.g., using indoor localization or navigation tools like Micello prevent and alert for accidents as falls and medical emergencies).

6.6 TOURISM SECTOR

Tourism is a sector with many different components and interrelated parts.[14] There are subsectors consisting of transport, accommodation, attractions, and activities that stimulate buying and thus the local business economy. If there is a data innovation for one such subsector, it must be common and applicable to all the subsectors.

6.6.1 Travel Arrangements

Tourism studies in academe mostly pay attention to sustainable development. Typically, sustainable development includes a strategic analysis, systematology, indicator systems of construction as well as appraisal, and correlation analysis of the interest groups. With regard to the environmental carrying capacity, and values vs. ethics, one of the more famous theories is the ecological footprint proposed by Canadian ecological economist William Rees in 1992. He demonstrated a quantitative calculation method expressed in productive land area or water area, whereby the goods and services are provided by the natural ecosystem, consumed by a certain population group, and then the waste is disposed by the natural ecosystem. These theoretical studies correspond to the practical tourism businesses and are captured in the rationality of travel arrangements. Good vacation planning can directly raise the occupancy of hotels and other facilities and bring in local revenue due to the buying habits of tourists, and indirectly eliminate any negative effects of tourism.

Generally, tourist packages of travel agencies are quite reasonably priced. However, many tourists consider such package deals too restrictive and try to self-arrange their vacations. Unfortunately, some such overconfident tourists, when they reach their destinations, often don't understand natural ecological boundaries or slip up by their ignorance of the activities they can partake in at their destinations. In other words, arranging travel without professional guidance can have a certain destructive

---

[14]http://en.wikiversity.org/wiki/Tourism/Introduction.

effect. Take a recent post, as an example, of a Chinese netizen criticizing a Xisha seafood journey. His was a self-service travel adventure organized by an outdoors club. In the post, he mentioned many prohibitive activities such as freely touching a nautilus and a *Tridacna sp.,* both marine organisms under first class protection.

Certainly, this does not mean prohibiting individual adventures. But travel arrangements require data as well as travel experience. Learning from others' experiences can help in arranging trips, and data mining of traveling tips, touring groups' itineraries, and landscape configurations can lead to improvements in the original itineraries and be made to conform to the tourists' original preferences.

6.6.2   Pushing Attractions

Hopper,[15] a start-up founded in 2007, opened access to its homepage in January 2014, a search engine that took a long time to "launch." Hopper CEO and the former Expedia engineers Frederic Lalonde's six years of secret preparations raised US$12 million of Series B from OMERS Ventures, Brightspark Ventures, and Atlas Venture. With additional funding in 2008 and 2011, the total financing amount was US$22 million. Some say it deals with the creation of a tourism search engine, but that is not entirely the case. Hopper main line is to push attractions based on place and an activity.

Jetpac and Tagwhat are also engaged in that kind of work. Jetpac specializes in travel pictures, over 750 million photos that are indexed from Facebook and were acquired by Google in August 2014 to improve the search for location information[16] based on identifying similar landmarks. Tagwhat is working on travel stories to make sure tourists don't miss great vacation spots and so using push services with users' geographical location.

6.6.3   Gourmet Food Recommendations

Gourmet food maybe is a highlight of tourism. For instance, in China, it is the fried sauce noodles of Beijing, the Goubuli stuffed buns of Tianjin, the marinated meat in baked buns of Xi'an, the Mapo Doufu of Chengdu, the salted duck of Nanjing, and the cross-the-bridge rice noodles of Yunnan; are all regional specialty foods. But people have their own food preferences. Food habits are well characterized by the proverb: "You are what you eat." It should be noted that data mining tools can help you acquire another type of trait for use in a recommending food. That is, to "Track and analyze what you eat."

Apart from food and recipe data, gourmet food recommenders need consumer behavioral data. Such data covers all offerings of existing gourmet food as well as reservation websites, such as browsing, searching, rating, commenting, adding into a shopping cart, deleting from a shopping cart, maintaining a wish list, participating in group purchasing, using coupons. There are additionally purchase returns as well

---

[15]http://techcrunch.com/2014/01/20/why-travel-startup-hopper-founded-in-2007-took-so-long-to-launch.
[16]http://techcrunch.com/2014/08/15/google-buys-jetpac-to-give-context-to-visual-searches.

as those consumers using the third-party websites doing price comparison, reading relevant evaluations, participating in discussions, exchanges on social media, and interactions with friends.

There are many data applications of food consumption. For instance, some consumers pay attention to food ingredients, so Food Genius can find food ingredients from 2,200 restaurant menus, and provide data services for both the upstream suppliers and consumers. Gojee recommends recipes that combine food ingredients consumers likely have on hand, and ties in auxiliary sale products. Other examples are Fooducate, FoodSmart, and Yelp, which deal in food purchases. They provide bar code scanning tools to help consumers acquire ingredients and include their nutritional values. They recommend where to buy cheaper materials, how to burn calories after eating, so as to encourage greater consumption. Then there are there the restaurant supporting services, such as OpenTable's food ordering service, NoWait's queue outside restaurant service, and Zagat's restaurant appraisal service.

6.6.4 Accommodation Bidding

Hotel reservations come with many options from online travel agencies, meta-search engines, and intermediaries' websites. According to a survey in Atmosphere Research (2012), on average, 95% of eventual guests have visited more than 22 reservations websites before making a final reservation.

Some booking websites have begun using various tricks. For instance, Getaroom and Orbitz hide prices, where the hotel brand and distribution channel is protected, thus their modus operandi is customer segmentation.

At present, big data bidding is accommodated by DealAngel, Guestmob, and Tripbirds. DealAngel is the first hotel deal "find" engine that helps consumers locate hotels priced below their fair market value, and plan their travel over special discount dates. Guestmob helps hotels solve the discount timing problem based on Price-Line patterns. Their specific reservation process consists of the following steps: (1) a consumer selects the destination, check-in time, and room type; (2) and then the corresponding group based on their accommodation needs; (3) Guestmob gives the deposit prices for hotels in this group; (4) the consumer then makes a payment. As for the hotel chosen in the group, Guestmob informs the consumer in advance (before the arrival time). Tripbirds mines the picture-sharing website with Instagram's data, and offers the consumers photos of real rooms instead of the official advertising pictures.

Of course, the essential goal in booking an accommodation is not finding the bottom-line hotel but rather a really cheap, decent, and comfortable hotel. Thus there is still a long way to go for data innovation in this subsector.

6.7 EDUCATION AND TRAINING SECTOR

From an industrial perspective, the education and training is a for profit sector. In fact the general trend worldwide is for education expenditure to take a large chunk of household consumption, and this includes preschool education, basic education, vocational education, interest expansion training, and professional certification training.

With the rapid growth of the Internet, the global e-learning market shows fast and significant increase. According to a GSV advisors' report, "the e-learning market globally from US$90bn to US$166.5 billion in 2015" will rise to "US$255 billion in 2017."[17] However, the field is still in disorderly competition at this early stage. There is presently little innovation, and speculative costs are low. Therefore it is worth exploring how to develop breakthrough innovations.

6.7.1 New Knowledge Appraisal Mechanism

Academic performance and cognitive abilities are the usual measures of a person's intelligence and capacity to acquire basic skills. However, these conventional appraisal methods of erudition and mental processes are is not always reliable. In many countries, there is valued, more or less, quantity over quality, especially when it comes to scholarly publications.

In October 2013, John Bohannon[18] at Harvard published an essay in *Science* that satirized sharply this phenomenon, entitled "Who's Afraid of Peer Review?" Bohannon, a biologist, concocted a spoof paper to expose the little or no scrutiny at many open-access journals. The spoof paper described anticancer properties of a chemical extracted from a lichen and was submitted 304 journals. To prevent writing style and fluency in English from revealing the author, Bohannon first translated the paper into French, using Google Translate, and retranslated the French back into English. Among the journals, 157 accepted the paper, 98 rejected, and 20 were still considering it. The average acceptance time was 40 days, and the average rejection time was 24 days. Many Indian and Japanese journals were among those fooled.

Except for the little native intelligence that can be scored using an intelligence test, the vast majority of what a person knows remains hidden. It's impossible to evaluate breadth of knowledge accurately by quantitative methods using small data statistics. We need more information as well as better technology to adequately appraise intelligence.

6.7.2 Innovative Continuing Education

Knowledge can, of course, go out of date. According to surveys conducted continually in the United States, among college graduates of the 1960s, the obsoleteness rate of what they studied in universities was 40% after 5 years, 65% after 10 years, and 75% after 15 years; for college graduates in 1976, the obsoleteness rate of what they studied in universities was up to 50% after 4 years and 100% after 10 years; for college graduates of the 1990s, what they studied in universities accounted for only 10% to 20% of what they learned in their entire lifetime, with the remaining 80% to 90% being supplemented or updated by job-related training.

---

[17]http://www.e-learningcentre.co.uk/resources/market_reports_/2012_17_market_predictions_from_gsv_advisors.
[18]http://www.sciencemag.org/content/342/6154/60.full.

Nowadays, people recognize the importance of continuing education. Many successful people use their spare time to further their education or training to attain higher professional positions. Nevertheless, this type of education (and training) is nothing more than cramming, and mostly to obtain compulsory credits for limited courses and routine examinations.

Continuing education should instead enable (1) lifelong learning, (2) individualized and personalized learning, (3) real-time learning through quizzes, (4) expansive on-demand learning through wide-ranging course selections, (5) crowdsourced open-courseware learning through computer applications, (6) peer-to-peer (P2P) interactive learning, (7) shared learning using people search, (8) shared learning through user-generated content (UGC), (9) immersion learning, and (10) anytime, anywhere learning. The way to implement all nine of these objectives is through data technologies. For instance, individualized and personalized learning depends on the implementation of a mining pattern by learners, whereas shared learning and UGC content sharing depend on graph mining and text semantic mining, respectively.

MOOC (massive open online courses) is a type of disruptive innovation in continuing education. MOOC's accomplishment is the entire mode of sharing that occurs with feedbacks on assignments, group discussions, problem set assessments, final examinations, and then certificates for those learners who desire them.

6.8   SERVICE SECTOR

The service sector is also known as a tertiary sector, in the three-sector theory offered by Allan Fisher in his 1935 book *The Clash of Progress and Security*. It's a sector covering a broad range of services, and generally, is regarded to be characterized by immateriality, nonstorage, and the simultaneity of production and consumption.

For the service sector, there are myriad classification standards, but there was no unified standard for a long time. Today, the standard categories are agreed to be general service, commercial service, and professional service. More specifically, (1) general service focus on individual and family basic activities (e.g., laundry, hairdressing, housekeeping, appliance repair, health and beauty, waste materials recovery) of residents in everyday life, (2) commercial service focus on socioeconomic activities (e.g., law, audit, accounting, tax, consulting, business advisory, market research), (3) professional service focus on activities related to scientific research, and professional development that are conducive to the generation, spreading, and application of knowledge.

The service sector depends on high human capital and technology in all three categories, and data innovations in the sector are changing the ethos of this sector from formerly pure individual accountability to genuine collective consensus.

6.8.1   Prolong Life: More Scientific

The desire for good health and longevity is the dream of many people. Modern fast-paced life, however, together with industrial pollution and pesticide residues makes us, more or less, susceptible to health problems. Different ways of preserving health, strengthening the body, and preventing disease so as to prolong a good life are admissible even if they lead ultimately to the reverse effect.

Theoretically, the logical way to prolong life is, first, to reject false promises. By mining different medical literature and pharmacopoeias, we should be able to eliminate false health information. Second is easy to self-screening for disease, as the saying goes "You are your own best doctor." So the logic in prolonging life is for each person commit to learning the literature with the benefit of big data. Admittedly, some symptoms can be confusing and not be caused by the same disease. Take joint pain as an example. There may be many explanations for it, such as a deficiency of the kidney, a strain at the joint, a recurrent injury, osteoarticular degenerative disease, osteoporosis, neck or lumbar vertebrae disease, gout, and other rheumatic diseases. Mining large-scale data to obtain adequate knowledge can help a person self-screen. Recently, a research paper on "identifying depression trend with social media data mining"[19] by the Harbin Institute of Technology (HIT) and Australia National University (ANU) started a heated discussion on the Internet. The data used for building a depression identification trend was from Sina Weibo. The model sifted through postings by over 200 patients with a major depressive disorder (MDD) and achieved an accuracy of 83%, as confirmed by the medical institutions. This group's posts were usually at about 23:00, and were more active in evening (about 30%), with keywords often including "death," "depressive order," "life," "agony," and "suicide"; 60% of these postings were from women. Third, promotion of personalized healthcare suited to the needs of an individual is the way to prolong life services and not product promotion.

6.8.2   Elderly Care: Technology-Enhanced, Enough?

Healthy aging is still a big challenge worldwide, along with the issues of healthcare. Technology-enhanced equipment, such as personal care robots, is not going to revolutionize elderly care as we know it; these are just attention-grabbing ideas. This is because such experimentations are at college campuses and removed from real life, they consider only the medical treatment and dwelling issues that leave the elderly feeling dissatisfied, and not the spiritual hunger and thirst of the neglected elderly.

Data innovation can improve life for the elderly, who spend lonely days and nights alone or in nursing homes. For instance, based on the data acquired from a sensor that can be embedded into the heel of a shoe, we can analyze "movement index" of the elderly and safeguard them, especially patients with Alzheimer's disease, instead of restricting their freedom. Other possibilities are (1) mining individual artistic interests, by offering college courses in calligraphy, painting, or dance; (2) reviewing personal health records within a certain jurisdiction to provide early warnings of low numbers of visitations from family.

---

[19]http://www.technology.org/2014/08/07/social-media-data-used-predict-depression-users

6.8.3 Legal Services: Occupational Changes

Occupations change over time and are not static. For example, data innovations have already led to great variations in the staffing of legal services, mainly affecting forensic staff, expert witnesses, judicial officers, and legal affairs staff.

1. *Forensic Staff and Lawyers in Investigating and Collecting Evidence.* In investigating and collecting evidence, forensic staff involvement is usually compulsory, and lawyers do the interviewing. The evidentiary materials obtained by lawyers are ineffectual unless verified in court through investigation by the forensic staff. In the future, algorithmers may be used to process evidence with criminal data and to immediately verify the evidence collected by forensic staff and from interviews by lawyers, and thus reduce the workload of both the prosecution and the defense as well as the court.

2. *Expert Witnesses for Serving the Court.* In court, the role of expert witnesses is to provide assistance in a case hearing based on their expertise in a certain subject. In the future, algorithmers may be used to verify the expert witnesses testimony with necessary data, or directly give objective opinions or conclusions as to the mining results regarding the case.

3. *Judicial Officers for Performing an Execution.* In an execution, what troubles judicial officers is the problem of (1) finding the person to be execution, (2) searching the property subjected to execution, (3) getting a person to assist with execution, and (4) handling the property to be executed. In the future, the algorithmers may be used to solve execution dilemmas by way of data mining. Data may even help resolve issues of invalid sanctions and inadequate investigations.

4. *Legal Affairs Staff for Sorting Case Files.* In the future, the algorithmers may be used to lessen the highly labor-intensive finishing work of cases, to reduce legal affairs staff's workload, and to accurately compute charges for lawyer services. Efficient algorithms may also be able to mine unstructured large-scale legal files or case data that is left idle, un-extracted, and un-used and then to display the results in a user friendly interface. Already there is Lex Machina,[20] which can capture all data of the Public Access Court Electronic Records, and can use natural language processing and legal text classification algorithms (both developed by Stanford University) to classify the cases, litigation excerpts, organization entities, patents, and outcomes of litigation for users' review and search. The same program could be used to measure the case win rates of law offices, or to evaluate the probability of litigation or reconciliation for a case.

6.8.4 Patents: The Maximum Open Data Resource

According to the World Intellectual Property Organization (WIPO), 90% to 95% of all the world's inventions can be found in patents. A patent is in fact the maximum

---

[20]http://techcrunch.com/2012/07/26/know-your-enemy-lex-machina-raises-2-million-for-ip-litigation-analytics.

open data resource in the world. When patent documents are used to assist R&D, the time required for research has been proved to be shortened by up to 60%, and cost lowered by as much as 40%.[21] This is because patents reveal the details of (1) technological development in an industry, (2) growth within an existing technology, (3) degree of competition within a technical field, (4) R&D of competitors, (5) applicable developing new technologies, as well as (6) potential market and economic value of new products.

Early patents can only be analyzed through references, or citations related to a patent documentation. Great Britain and America have long established special citation databanks. The earliest is the British Patents Citation Index (PCI) established in 1995, the predecessor of Derwent Innovations Index (DII) that was built in 1999 jointly with the American Institute for Scientific Information (ISI) with the data dating back to 1963 and covering 18 million patents from more than 40 patent institutions across the world. This is an important patent information organization, which, by the quantitative statistics of the citation relations among the patent literature, contains insights into the R&D state and technological levels of different technological fields.

No doubt, more and more people are realizing that data mining tools and data visualization methods should be used to analyze patent resources. For example, Aureka's topographic map and multilevel citation tree can help us search and find the development stages of patents and development tracks of industries. Of course, mining of patent data apart from citations could obtain other useful information and knowledge, as is left to future data innovations as well as industrialization.

6.8.5 Meteorological Data Services: How to Commercialize?

Meteorology is the interdisciplinary scientific study of the atmosphere that concentrates on atmospheric waves as well as their variations under different weather conditions. Current research methods, such as observational studies, theoretical research, numerical modeling research, and experimental studies involve the acquisition, observation, management, and analysis of meteorological data. Therefore the service offered in this professional service sector is weather forecasts – a meteorological data service.

Weather forecasts may actually be a special type of "monopoly." Currently, just about worldwide, laws and regulations stipulate that except for "the weather stations of the competent state meteorological departments at various levels, … any other organization or individual must not issue public meteorological forecasts and severe weather warnings." Thus to enhance the product quality of a weather forecast at this stage requires the meteorological departments themselves to transform to data industry thinking and adopt new techniques like data mining.

Yet, today, a few corporations are beginning to seek commercial value from meteorological data analysis. Instead of the quoted results of weather forecasts, Merck[22] is using meteorological data to forecast allergy outbreak months in different zones (by

---

[21]http://www.itc.gov.hk/en/doc/consultation/consultation_paper/companies/P99.pdf.
[22]http://www.grtcorp.com/content/big-data-puts-weather-work-businesses.

zip code), and then to sell its allergy medicine; Claritin is featured in joint promotions with Walmart. Sears Roebuck stocks its warehouse using real-time monitoring of meteorological data, so as to have sufficient quantities of snow blowers prior to a snowstorm and to ensure that enough air conditioners are in stock when temperatures climb to severe highs. AECOM has made it marketing policy to have representative seasonal goods ready based on meteorological data, among these an anti-static spray used only in conditions of low humidity. DHL adjusts its daily 3,000 cargo flight schedules based on meteorological data to an accuracy of one minute. Liberty Mutual Group identifies false claims using meteorological data, such as in deciding through association analysis whether a hailstorm was the cause of building damage, based on hail size and intensity and the building's former condition.

To this end, meteorological data services are not limited to weather forecasts. Apparently, restrictions of laws and regulations can be circumvented with professional data products that intersect with users' particular needs.

6.9 MEDIA, SPORTS, AND THE ENTERTAINMENT SECTOR

The media, sports, and entertainment sector is a paragon of cross-border integration of digital media in the data industry. Many practitioners of this sector are engaged in data innovations, consciously or not. For example, in the United States, the government acknowledged, in 2006, that paper-based circulation and electronic READ are of equal value. Big data analysis is already being used in every link of the film industry, and all virtual equipment and items in online games are fittingly included as well.

6.9.1 Data Talent Scout

Formerly, actors and other performers were selected by talent scouts, and more recently, from TV talent shows. Talent scouts, usually belonging to or signed with talent agencies, provided a professional perspective for people with star potential. Britney Spears and N'sync were discovered this way. Reality TV shows have been popular now for more than a decade. One example is *American Idol* created by Simon Cowell, and launched by the FOX network. *American Idol*'s first season led to the discovery of Kelly Clarkson, who later won a Grammy Award. Needless to say, as in sports, talent discovery depends on a talent scout's intuitive judgment.

Michael Lewis's 2011 book *Moneyball: The Art of Winning an Unfair Game* describes a real life instance of data selecting a ballplayer. The book was rated as one of the seventy-five must-buy business books by *Fortune*, and it is regarded as a great book on contrarian investment. In the book, baseball talent scouts demonstrate that accurate data analysis has precedence over intuitive judgment or compulsion [1]. Although the analysis method used for this book is trivial, it remains in contention a new impartial template.

At present, about 20 NBA teams are using data mining tools – primarily Advanced Scout, developed by IBM for tactical adjustments. The players are asked to wear devices that (1) monitor heart rate, respiration, speed, or running distance, and then

(2) defects in the starting lineup or tactical layouts are analyzed as per "full court pressure." Based on the results, the players receive (3) personalized training or physical rehabilitation treatment, and coaches (4) work out reasonable strategies of defense or attack to guide the ballplayers. The National Hockey League uses a similar data mining application NHL-ICE that can be accessed in real time by the coaches, speakers, journalists, and fans.

6.9.2   Interactive Script

The first "Easter egg" is an interactive work introduced in 1977 by game designer Warren Robinett in *Atari's Adventure*, in the early graphical version of the game. In the game an Easter egg is often accompanied by a series secret response that "occurs as a result of an undocumented set of commands."[23] For example, Google's search responds to "Do A Barrel Roll"[24] in the search box by tilting the page 360 degrees, as if a pilot were maneuvering an aircraft. Most gamers like the Easter egg's surprise and have named it an extension task.

A decade ago, the same concept of "Easter egg" began to invade film and television circles. The idea was to increase audience participation, including in the selection of subject matter, script, director, and actor. This changed the traditional setup of media being controlled by a small industry group and being passively accepted by the public. At present, such audience interaction is being fundamentally transformed into an accumulation of big data. For instance, before making TV series the *House of Cards*, Netflix mined data of nearly 3,000 users on its DVD and online video rental site, including three million search queries, four million video appraisals, and mouse pauses as well as position information. The query results returned matches for director David Fincher, actor Kevin Spacey, and a 1990 British political TV show of the same name. Another example was an Emmy Award to YouTube for a video recommendation algorithm that "makes the site feel deeply personal and combs through a massive collection of content to find the videos that will capture your attention for a few minutes longer."[25]

However, the real interactive script lies not in script customization but rather in script modification, that is, the movie or TV show with the plot based on audience preferences in setting multiple branch points. This is not impossible. Hollywood has explored this idea because of underperformances at the box office. The idea came from a former Wall Street analyst Alexis Kirke[26] who in early 2013 released an interactive film called *Many Worlds* that immersed the audience in the plot. Kirke's objective was to make the movie "branch off into multiple divergent story lines when activated." For the technology to work, the audience members agreed to wear sensors, for monitoring heart rate, tracking brain waves, registering perspiration levels, and measuring muscle tension. Based on different sensory responses, the audiences

---

[23]http://en.wikipedia.org/wiki/Easter_egg_%28interaction_design%29.
[24]http://content.usatoday.com/communities/gamehunters/post/2011/12/googles-let-it-snow-feature-makes-web-winter-wonderland-/1#.VWhxwo6qqko.
[25]http://www.theverge.com/2013/8/1/4578544/youtube-wins-first-emmy-for-video-recommendations-you-cant-resist.
[26]http://www.bbc.com/news/technology-21429437.

could change the narrative structure or direction of the film. For example, if some-one "started to get tense, the film flicked to a more relaxing sequence"; if another felt "encouraged to relax, the footage at the next branch point switched to a dramatic moment with an undercurrent of expected violence."

Similar types of explorations exist in online game scripts, for the beginning, in the way the various editions are set up. The first minutes or hours of a game are very important and can draw in or turn away a large number of gamers. Besides the usual considerations of usability, elegance of interface, the validity of a free trial, the learning curve, and the tutorial quality, there is user performance that is expected to improve with gaming experiences before users quit playing a game.

6.10   PUBLIC SECTOR

The public sector is the part of the economy that provides various government ser-vices.[27] These services can be fulfilled by departments of government, nonprofit organizations, and nongovernmental organizations concerned with particular issues of public interest.

Public service management can be classified into four categories based on content and form: (1) basic public services such as water, electricity, gas, and communica-tions; (2) economic public services such as advancements in science and technology industries; (3) social public services such as public education, medical treatment, and social welfare; and (4) public security services such as firefighting and policing.

Since in one way or another public administration extends to all members of soci-ety, big data has the potential to make all these management functions more efficient and save taxpayer money. In addition, as discussed in Chapter 5, data innovation can help enormously in the administrative work of the police and fire departments.

6.10.1   Wargaming

Wargaming is a simulation of different military operations. Early wargaming can be traced the Warring States period of China (475–221 BC) when the ancients used peb-bles and twigs to demonstrate military activities and maneuvers and evaded the need for actual hostilities. In 1811 Baron von Reisswitz, a civilian war counselor to the Prussian court, invented modern wargaming, which consists of a map and several chess pieces under a set of rules. Wargaming has changed dramatically to contem-porary computer-assisted wargames as information technology has developed. These computer wargames involve a number of combat troops, an organizational structure, a weapons, and tactics, along with a set of emulation rules for the command center, combat training center, and implementation units in order to demonstrate specific bat-tle processes, around major combat readiness issues, according to various strategies and tactics under simulated combat circumstances.

However, there are a lot of wargaming of disputes. This is because, first, wargam-ing relies too much on relevant software and lacks software of existing campaign-level

[27]http://en.wikipedia.org/wiki/Public_sector.

or above; second, computers do massive calculations only as a supplementary service, that is, a large number of rules on warfare and on command and control simulation are conducted by manual operations; third, it is very difficult to compare wargame results with real outcomes of war by conventional methods using statistics and operations research.

To break through the existing impasses, wargaming has to solve two problems. One is to obtain more data resources, including meteorological data, remote sensing data, high-altitude unmanned aerial vehicles (UAV), as well as light aircraft probe data, that is, apart from the military data of both sides such as tactical rules, war strength, weapons effects, and trench terrain. The other is to directly use data mining tools, on the one hand, to find tactics or rules that cannot be found utilizing traditional analysis tools, and, on the other hand, to assess the wargame results in terms of campaign course and real outcomes of war, instead of the repeatedly labor-intensive work of "designing software based on demand, and then simulating exercise with application." With data mining tools, future military counterwork may actually be decided by virtual wargaming, which is why this example has importance.

6.10.2 Public Opinion Analysis

Public opinion analysis refer to monitoring the evolution of public appeals as to topics of natural disasters, worker safety and injustices, public health, public authority misconduct, judicial questions, economic sanctions affecting citizen livelihoods, and social ideological movements.

In cyberspace, public opinions spread faster and have a wider influence and evade real-time monitoring or police intervention. In cyberspace there are urgent demands for emergency disposal, public communication, and public opinion guidance. This is worth considering whether new data tools can be used to examine and predict public opinion trends. Blab,[28] the "crystal ball" of social media, based on the public opinion trends, can predict public opinion hotspots one to three days in advance, help government and enterprises address rifts and crises. Blab can monitor the nearly fifty thousand data sources from various blogs and news websites, including Facebook, Twitter, and Youtube, and it then analyzes the message propagation paths and patterns to forecast the time of a next similar hotspot. Particularly useful to law enforcement is that Blab can measure weight of the message source and locate the core or pivotal person behind the publicly transmitted opinion.

---

[28]http://www.cnn.com/2014/09/03/business/blab-crystal-ball-social-media/index.html.

7

BUSINESS MODELS IN THE DATA INDUSTRY

Peter Drucker, a widely acclaimed management theorist, makes the important point in his 1999 book *Management* that from the perspective of the enterprise, the competition between enterprises is "not the competition of products but the competition of business model [56]." That is to say, from the perspective of an industry, the way a business model fares among competing business models can increase (or decrease) its industry profits and even create changes in the industry. Such changes in traditional industries often allow emerging industries opportunities to develop. Business model innovation, for the purpose of deploying an original business model, has been especially important in the development of the emerging data industry.

7.1 GENERAL ANALYSIS OF THE BUSINESS MODEL

In defining what constitutes a business model, scholars have been keen on elaborating the annotation and classification of a business model. Since the 1950s, there have been hundreds of such definitions. The most cited was suggested in 1998 by Paul Timmers [57] who defined a business model as "an architecture for the product, service and information flows, including a description of the various business actors and their roles; a description of the potential benefits for the various business actors; and a description of the sources of revenues." The most widely accepted business model definition today is that from *Clarifying Business Models: Origins, Present, and Future of the Concept* by Alexander Osterwalder [58] published in 2005. Osterwalder defines the business model to be "a conceptual tool that contains a set of

The Data Industry: The Business and Economics of Information and Big Data, First Edition. Chunlei Tang.
© 2016 John Wiley & Sons, Inc. Published 2016 by John Wiley & Sons, Inc.

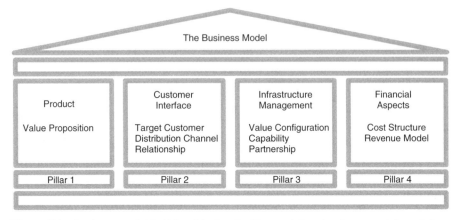

Figure 7.1   Business model building blocks: 4 pillars and 9 main elements. Adapted from [58]

elements and their relationships and allows expressing the business logic of a specific firm." In brief, it is generally agreed that a business model is a set of elements formed by a logical plan and a purpose to create and drive value.

7.1.1   A Set of Elements and Their Relationships

The elements of a business model can be diverse. In 2003 Michael Morris and colleagues [59] at Syracuse University published an article, in the *Journal of Business Research*, in which they compared business model definitions of over thirty scholars from 1996 to 2002, and among them, one by Gray Hamel a foremost master of strategy who won the McKinsey Award[1] several times. They found that there can be as many as twenty-five elements in a business model, that the average number of definitions for each element is between three and eight, and that some definitions are repeated such as the value proposition, partnership, product, internal infrastructure, target market, resources, and capacity. Subsequent to this study, in 2005 Osterwalder [60] brought the number down to nine main elements of the business model as used in academic circles (i.e., value proposition, target customer, distribution channel, relationship, value configuration, capability, partnership, cost structure, and revenue model) and sorted them by product, customer interface, infrastructure management, and financial aspects into four pillars (shown in Fig. 7.1).

As can be gathered from these elements, the business model is mainly about maximizing (1) customer value, (2) business portfolio, (3) efficiency, (4) support systems, (5) profitability, (6) fulfillment methods, (7) core expertise, and (8) overall organization. Of these, for the business portfolio, high efficiency and having good support systems are the critical preconditions; core expertise relates to the competency that

---

[1]McKinsey Award was founded in 1959 jointly by McKinsey Foundation for Management Research and Harvard Business Review, aiming to commend the authors of two best articles published on Harvard Business Review.

is critical to drawing revenue; and maximizing customer value is the more subjective pursuit of profitability, which is the objective result.

7.1.2 Forming a Specific Business Logic

The business model follows essentially a logic of transverse connections and vertical integrations of elements relating to product design, customer interface, infrastructure management, and financial oversight. By the same logic, due to Morris and Oster-walder, the elements can be further reordered and combined to communicate a core business plan along three lines: First is *economic logic,* which means preparing a logical statement of the enterprise's vision for acquiring and sustaining its profit stream over time [60]. The main elements include the cost structure and revenue model, and then other elements such as the profit model, pricing strategy, and output optimization. Second is *operation logic*, which is a statement on infrastructure management, including relations between business income and the industry value chain, but excluding the market, environment, and stakeholders. The main elements include core capabilities and partnership arrangements, while other elements include delivery mode, management flow (e.g., resource flow, business flow, logistics flow), and knowledge management. Third is *strategy logic,* which is formed by preparing a statement on the development of product and customer interfaces, along with a description of industrial interests from market participants. The main elements include the value proposition, target customer, distribution channel, and relationship, and other elements include value creation, differentiation, and vision.

These statements express the special purpose a business has with relation to the external economy and its projected operations based on other businesses already operating within the industry. For interindustry objectives, the emphasis is on solving problems in the market or meeting unsatisfied customer demand, which is different from an intraindustry objective regarding revenue sources or profits.

7.1.3 Creating and Commercializing Value

In a business model, value not only refers to the profits but also to a process for finding, creating, and acquiring value. Value investigation, creation, and acquisition are simultaneously manifested in the value demanded by the target customer, the value intended in the products, the value behind the creation of the business operation, the value delivered to the distribution channel, and the value protected by strategic control. These four ways were proposed in 2001 by Raphael Amit and colleagues [61] of the Wharton School at the University of Pennsylvania. Included were also novelty, lock-in, complementarities, and efficiency (shown in Fig. 7.2).

With respect to value, a good business model may have the absurdity of including both reproducibility and barriers to competition. On the one hand, reproducibility, means the product is easy to imitate or to adapt, for even plagiarism and micro-innovation are recognized for their originalities. But it is inadvisable to attach too much importance to logic and neglect value, or to give too much emphasis to generating profits. The other hand, for value protection, there must be created a barrier to competition. This can take the form of balancing value creation with the

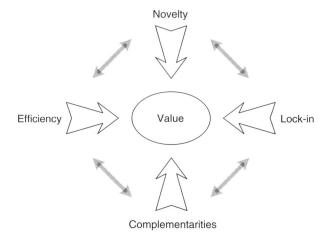

Figure 7.2   Ways of value creation and acquisition. Adapted from [61]

acquisition of a trade standard, leader status, brand image, industry scale, customer loyalty, patents, or other exclusive rights, so that other competitors find it hard to mechanically copy.

7.2   DATA INDUSTRY BUSINESS MODELS

Throughout this book, to describe the data industry business models, we have employed a conceptualization that consists of three dimensions: resources, technology, and capital. But we find it necessary to extract and add some insights from different industries that can enrich the economic profile of the data industry.

7.2.1   A Resource-Based View: Resource Possession

A resource industry is, by definition, restricted by the resources available. Take the example of oil and natural gas. There is only so much oil and natural gas available below the earth. From a resource-based view (RBV), the business model of resource industries is focused on various tangible and intangible nonrenewable natural resources, and therefore called a resource proprietorship type of the business model. This is obvious with respect to (1) the core competitive capability advantaged by resources, (2) the value proposition highly depending on the amount of resources, and (3) the relatively weak elements that are in customer interface and infrastructure management. From the perspective of forming a strategic plan, it is based on the utilization of unique resource models, and from a value perspective, it is characterized by possession of certain prized resources.

The resource proprietorship type of the business model can further be subdivided into resource utilization, resource creation, resource grafting, and resource integration. In general, this business model gives priority to resource utilization;

technical advancement and division refining, along with uncomplicated resource utilization, can be combined into resource creation (e.g., applying a purification process to improve the resource), resource grafting (e.g., joining a channel to increase resource output), or resource integration (e.g., combining with another resource to extend profit margins for both resources).

Some features of the business model of resource industries can be applied to the data industry business model. But, first, there are two differences to consider: high pollution levels and the non-renewability of resource industries. Second, the data industry is highly dependent on data resources as well, but some such elements like core capacity and value proposition are more among the advantages of data resources. Of course, there are the physical differences too.[2] Data resources require hardware devices, a storage medium and transmission equipment. There are no geographical site considerations for embeddedness, resource richness, or supply convenience.

For the data industry there are further different proprietorial modes that may be necessary to bypass in the task of business model extension. One is *domain control,* which is essentially exclusive possession of data resources within a field, and generally this entails specialized control such as monopoly within the financial sector or sole expertise in a medical field. The elements (e.g., customer target, relationship, partnership) that are enhanced in these domains combine to form exclusiveness. Its logic formation is under strategic control. Its value is characterized by occupancy of the industrial value chain. The other is *industry support,* which is based on shared possession of data resources outside the industry. Where the business model intercedes is in value creation by the supporting enterprises. Here the elements (e.g., value configuration, distribution channel) are based on brand transformation initiated on a basis of customer demand. Its logic formation belongs to operation improvement. Its value is carried out through value implementation.

7.2.2   A Dynamic-Capability View: Endogenous Capacity

From a dynamic-capability view (DCV), the "endogenous capacity" type is another common business model in traditional industries. What constitutes a "capability"? Richard Makadok, at Emory University's Goizueta Business School, has emphasized the distinction between capabilities and resources. As defined by Makadok, capability is "a special type of resource, specifically an organizationally embedded non-transferable firm-specific resource whose purpose is to improve the productivity of the other resources possessed by the firm [62]." Thus the objective of an "endogenous capacity" type of business model is to collect and cultivate internally means of sustaining innovation. Among the elements cited for the business model, this is manifested in (1) the core competitive capability due to technological specialization; (2) the cost structure in which human resources and knowledge are applied in higher proportion; and (3) the value proposition and distribution channel where diversification and multidimensionality are most essential. From the perspective of business strategy, endogenous capacity relates to the operational structure and includes the profit

---

[2]To simplify, we use "data enterprises" as a abbreviation form that refers to "enterprises of the data industry."

margin; from the perspective of value, it relates to the newness and efficient operation of the business innovation.

Endogenous capacity from a weak to a strong development course can be divided into technology leading, sustaining accumulation, brand licensing, division, and interaction. In its early growth phases, this business model generally gives precedence to technological breakthroughs, based on the advantages of technological specialization. Sustaining operation for a period of time in order to continuously accumulate value can ensure stability. After a period of accumulation, there may be added more specialized development and increased scale, a contracting mode of brand licensing for sharing brand resources to increase profits, or an exchange of mutual interests to acquire a new profit resources.

Differences in "endogenous capacity" type between the data industry and traditional industries include (1) differences in support of core capacities, in the Art + Technology initiative (i.e., creative content and data technology), no longer emphasizes technical breakthroughs and optimization; (2) differences in the cost structure in that benefits from open source, social collaborations, and crowdsourcing can save on expenditures in human resources, knowledge assets, and other aspects of costs; (3) differences in operation modes from purely oriented by finance or achievement, and based on an industrial value chain, to multilevel promotion and integration based on operation and the strategy of market segmentation; and (4) differences in value acquisition in that novelty or efficiency is no longer the singular objective of the industrial chain, and this greatly increases the complementary values among the different industrial chains.

For these reasons, the data industry has embedded in its business models two types of extensions. The first is *crowdsourcing*. Crowdsourcing is, for the purpose of cost reduction, a new effective way to use, via the Internet, volunteer employees who have interest and are capable of spending leisure time to make a small amount of earnings. These partnerships, or informal relationships, reduce costs, and thus the logic behind this arrangement. Crowdsourcing comes under operation control, and its value suggests complementarity. The second is *customized products*. Customization is based on market segmentation; for instance, the market can be mass, niche, segmented, diversified, or multi-sided, and the product differentiation takes shape by meeting or catering to the demands of individual customers or segmented customer groups. With regard to the data business model elements, data innovation is used for raising the intensity of customer targeting and relationships. With regard to its logic formation, it is integration that is characterized by both novelty and value complementarity.

7.2.3 A Capital-Based View: Venture-Capital Operation

The venture capital gained through mergers, acquisitions, divestitures, and alliances can play an important role in developing emerging industries. From a capital-based view (CBV), the "venture-capital operation" type of the business model is characterized by utilizing costs or market rules to realize increments of value. In regard to elements, it is manifested in (1) the value configuration and capability, which are of the means by which the product can attract venture capital input, and (2) the distribution channel, where the asset transactions (e.g., buyout, lease, mortgage, pledge, partition, buyback) take place. From the perspective of logic forming, acquired venture capital relates to the strategic aspect of seeking investment and reinvestment opportunities in external markets. From a value perspective, it is characterized by venture capital occupation and complementarity.

In the business model, the cyclical pattern of venture capital expansion, integration, and contraction may take the form of business incubation, equity and debt investments, additional investment, capital lease, fund operation, transaction of property rights, enterprise merging, assets stripping, corporate separation or spin-off, and share repurchasing. However, there is a big difference in defining, quantifying, and evaluating data assets in the "venture-capital operation" type of data industry business models. In this regard, it is better to lay out fair, equitable and statutory regulations or to verify, valuate, and report data assets by applying appropriate methods so as to offer a basis for measuring value. The same intangible assets (e.g., patent, trademark, brand, land-use right, franchise) offer a certain basis and prerequisites for change of title, equity innovation or liquidation, and auction, thereby promoting, growing, and developing the data industry.

7.3 INNOVATION OF DATA INDUSTRY BUSINESS MODELS

The business model is not static. However, its capacity of value acquisition will weaken and eventually disappear due to market competition, technical breakthroughs, or changes in customer demand. Therefore timely discarding and updating of business models is critical.

Business models require active and dynamic adjustments from time to time to keep them profitable, along with value lock-ins and technological advances. With respect to objects, tasks, and functions, updates may be needed to the list of elements, logic formation, and the value proposal, likely to include new products, customer interface innovations, infrastructure management reforms, financial adjustments, and strategic resources reorganizations and acquisition optimizations.

To stay innovative, the data industry business models must be continually (1) seeking industrial development of the data industry by exploring new value creation and acquisition; (2) learning from challenges of traditional industries undergoing industrial upgrading and transformation. Understanding basic rules and methods of business model innovation is a precondition for creating a competitive advantage in overall industrial achievements within a region and even countries.

7.3.1 Sources

The seven sources of innovation are, as listed by Peter Drucker in his book *Innovation and Entrepreneurship* [63]: the unexpected, the incongruity, process need, industry and market structure change, demographics, changes in perception and mood, and new knowledge. In 1988 another source [64] was added by Eric von Hippel at the MIT Sloan School of Management: user innovation. The meaning of "user" in this context refers to enterprises or individual customers who innovate in their use of a product or service become for the industry sources of innovation themselves.

In binding together these essential characteristics of the data industry – data resource possession, data asset quantification, data technology research and development, data industrial funds investment, and customer orientation – we can expect that to be innovative, data industry business models should have the following characteristics:

1. *The Unexpected in Data Resource Possession.* The unexpected implying "opportunity," includes unexpected success or failure. Even though an industrial link can allocate value and enlarge a value-added space in the data industry chain, the occurrence of an unexpected failure in the industrial chain or unexpected success in a periphery industrial chain can indirect affect the data resources. In the example of the taxi industry planning its transportation data resources, its operation's value is easily encroached by a rival taxi booking mobile application (e.g., Uber). The taxi business model has made no allowance for data innovations; all it can do is simply open up its communication channels with the drivers and passengers.

2. *Changes in Perception and Mood, Process Needs in Data Asset Quantification.* Given "a half glass of water," different people may perceive it as "half full" or "half empty." When cognition changes, its meanings also change and the change of the fact itself is ignored. It should be noted that such changes can be inspected, utilized, and redefined. The data value recognition, namely that the considered data is a core asset, is vital during the development course of the data industry, as changes in perception and mood and process needs can directly affect innovative course of data industry business models. People who can seize the opportunity, which is to recognize the asset value of the data before others can, more easily secure the data industry profits.

3. *New Knowledge in Data Technology Research and Development.* Data mining is deep knowledge discovery through data inference and exploration. From this perspective, data technology research and development belongs to *new forms of knowledge*, and that is its endogenous core innovative capacity.

4. *Incongruity, Industry and Market Structure Change, in Data Industrial Funds Investment.* Timely investment of industrial capital means seizing opportunities as incongruities occur changing industry and market structures. These generally include (1) an incongruity between industry assumptions and market expectations that is manifest in that market expectations misleading by the profits obtained from existing data innovations miscalculated the future value abundance of the entire data industry chain, and then the pace of data industry development will slow; (2) an incongruity between product demand and industrial profit that is manifest in that demand for data products provided by the data industry on the market is growing, but the profit margin does not corresponding to the increase resulting in startups taking a hit in terms of confidence; and (3) an endogenous incongruities of industry that is manifest in fault a between laboratory R&D and market applications of data products.

5. *Demographics, User Innovation in Customer Orientation.*   Businesses in a market economy survive by producing products that customers are willing and able to purchase. For this reason enterprises focus their activities and products on customer demand. Two innovation measures that should be added are demographic and user innovation aimed at customer demand, in which the demographic is used for identifying a market and segmenting the customers and user innovation is used for describing consumer tastes or preferences in adjustments they make to products. There is further, in comparison with traditional statistical methods, data mining because this is used to support customer demand and preferences mining to find, create, and acquire value in a wider, more convenient, and reliable way.

7.3.2   Methods

The thinking or mode of business model innovation is much less important than the method. Using the correct method is conducive to questioning, challenging, and converting old business models as well as to inventing, designing, and realizing novel approaches to the business model.

With regard to method, the conventional approach to innovation is creativity stimulation. Included are various types of inventive problem solutions, scenario analyses, trend analyses, roadmap methods, quality function deployments, and other creative technique such as SWOT (i.e., strengths, weaknesses, opportunities, threats) analyses, brainstorming, and the seven-step problem solving techniques. However, besides creative ideas, business model innovation should include the industry or organization itself, the existing and potential competitors; there may be ways to estimate products, observe customers, product prototypes, and deduct scenarios as well. Altogether, this is a system design process similar to a strategic blueprint.

Therefore, with regard to innovative data industry business models, it is not simply a matter of selecting or more comprehensively using traditional methods; success depends on combining sources of innovation in the entire value chain that is characteristics of the data industry.

1. *Innovation in Total Element.*   As can be seen from the general analysis above, a number of elements are indispensable to creating a business plan. Categories, quantity, and hierarchical division of these elements should be discussed and a consensus reached as to how they work together in the realization of the total element innovation. Total element innovation does not only refer to innovation of products, customer interface, basic management, and profit making, it also refers to adapting the contributions to innovation of all elements. Every category of element innovations should be involved any adjustments in the system, organization, culture, strategy, and market, so that result is collaborative, on the one hand, in creating an innovative mood within the enterprise, and, on the other hand, to justify readjusting or reallocating resources from other elements so as to promote the innovation of a distinct element.

   Free of charge is a kind of innovative approach to the pricing element. Free of charge includes advertising-based totally free mode, basically free mode in "freemium", and use of free products to draw in repetitive consumption via "bait and hook" mode. Such a single element innovation will call for collaborative innovation, for instance, in cost structure changes from cost driven to value driven, in distribution channel transmission values to customers, and in customer segmentations. Conversely, one could begin with the integration and allocation of resources to value proposition, distribution channel, value configuration, and cost structure so that the overall change is conducive to the element innovation of each group or customer relation in target consumers, that is, value demand, value proposition, value creation, value transmission, and value protection.

2. *Innovation in the Entire Value Chain.* From the perspective of Porter's value chain, the opportunities of business model innovation exist in about five stages. In the stages of value demand and value proposition, it is necessary to master the market trend, customer preferences, and competition flow. The stage of value creation begins from technological innovation with product innovation as its core and based on technological research for the product and process development. The stage of value transmission is at marketing innovation mode. The stage of value protection is at beating the competition by exclusive innovation.

With regard to the data industry, innovation of value demand and value proposition is realized by mining consumer preferences or market demand. The innovation of value creation is decided by the data product design. The innovation of value transmission is during data marketing, and the innovation of value protection is implemented by monopolizing data resources.

7.3.3   A Paradox

As an extension of Schumpeter's gale[3] (i.e., creative destruction), disruptive innovation [65] was first proposed in 1997 by Clayton Christensen of the Harvard Business School. All existing or future data innovations listed above have proved that data technology is a disruptive technology, and it is critical to the future development of the data industry as an emerging industry.

However, as believed by Henry Chesbrough, who coined the term "open innovation," "Technology by itself has no single objective value. The economic value of a technology remains latent until it is commercialized in some way via a business model. In fact, it is probably true that a mediocre technology pursued within a great business model may be more valuable that a great technology exploited via a mediocre business model [66]." It is important to recognize that as a disruptive technology, the advancement of data technology is often beyond existing levels of public awareness and the degree of acceptance. Thus it is relatively difficult to implement in the innovation of data industry business models. On the one hand, data resource providers distrust new data technologies, in particular due to risk and uncertainty. On the other hand, some executives will strongly suppress a new business model due to the disruptive innovative challenges to existing products, industries, and even

[3]http://en.wikipedia.org/wiki/Creative_destruction.

systems. Hence it would be ideal if those charged with innovation would fully cooperate in all political, academic, and industrial circles to build bigger and better industrial development models, so as to avoid the innovation paradox between the data technology and business model.

8

OPERATING MODELS IN THE DATA INDUSTRY

Ideally, as should be clear from the preceding chapter, the business model is conceived to generate winner-take-all effects. But, although a large number of enterprises have continued to innovate their business models, and with some success, a truly innovative business model is hard to come by, and is not as easily defensible as it once was. Enterprises are always on the lookout for new forms of competitive advantages. The answer may be found in the "operating" model,[1] delineated by management guru Peter Drucker in his 1954 book *The Practice of Management*.

An operating model describes how to effectively combine such proprietary resources as human capital, financial capital, and material capital to keep up a continuous flow of profits. Unlike the business model, the operating model reflects the business philosophy, management style, and force field assessment of an enterprise, all of which vary depending on the enterprise. Competitive advantages are obtained through innovations in the operating model, as analyzed based on Porter's five forces. An enterprise's competitive advantage is a distinctiveness that is enduring and hard to copy directly, and so decides the profitability of the enterprise by affecting price, cost, and the necessary investment by the enterprise. In fact the combined rationalities of the operating models in the data industry are the major factor behind the development of the data industry.

---

[1] http://sloanreview.mit.edu/article/what-is-your-management-model.

The Data Industry: The Business and Economics of Information and Big Data, First Edition. Chunlei Tang.
© 2016 John Wiley & Sons, Inc. Published 2016 by John Wiley & Sons, Inc.

8.1   GENERAL ANALYSIS OF THE OPERATING MODEL

Henri Fayol, a founder of modern management methods, maintained that management should consist of[2] forecasting, planning, organizing, commanding, coordinating, and controlling. The operating model, from this standpoint, involves choices made by the enterprise, and in accord with its business philosophy, they define objectives, motivate labor effort, coordinate production activities, and allocate resources aimed at a particular market. These choices shape the enterprise's practices and make it possible to further optimize its business model.

8.1.1   Strategic Management

Prior to the 1950s, the term "strategy" was primarily used for war and politics, and not by businesses. In 1957, Igor Ansoff, who pioneered the concept of strategic management, introduced the term "strategy" into the field of economics in "Strategy of Diversification," published in the *Harvard Business Review*. In 1987, Ansoff's four marketing plan elements were added another element by Henry Mintzberg of McGill University to make it a 5P strategy model. This was also when the term "strategy" was explicitly recognized to have multiple definitions, namely plan, ploy, pattern, position, and perspective.

Later contributors[3] summarized and found the 5P a convenient way[4] to classify strategy in terms of levels, from top to bottom. Generally, corporate-level strategy is responsible for the market definition, business-level strategy is responsible for market navigation, and functional-level strategy is the foundation that supports both corporate-level and business-level strategies. All in all, (1) the corporate-level strategy is the strategic master plan of an individual enterprise, and has to consider the scope and focus of business activities according to its business philosophy and then to set priorities in allocating resources to functional areas of the enterprise; (2) the business-level strategy addresses the competitive advantage to be gained, by matching the relationships of products with consumer demand and competitor analyses; (3) the functional-level strategy does the coordinating of the functional areas of the organization (e.g., marketing, finance, human resources, production, and research and development) so that each functional area upholds and contributes to business-level or corporate-level strategies.

From the perspective of strategic management, in designing or choosing an operating model, the enterprise is strategically prepared to manage within a dynamically changing environment and to surmount any severe challenges to its capital, market, production, research and development, human resources, and public relations, and moreover to take full advantage of available opportunities and create new opportunities.

---

[2]http://en.wikipedia.org/wiki/Management.

[3]Among the numerous latter contributors, the most influential were Michael Porter, Henry Mintzberg, Jim Collins, etc.

[4]http://www.referenceforbusiness.com/management/Sc-Str/Strategy-Levels.html.

8.1.2   Competitiveness

Competitiveness, which abides by a set of relative indicators, is more difficult to accurately measure. Competition can only be evaluated accurately when more than two participants have similar outlooks. However, because there are different competitiveness theories, measuring competitiveness narrows down to weighing the suitability of a certain objective. Take core competency[5] as an example. The objective is a "harmonized combination of multiple resources and skills" of an individual enterprise, and the yardstick is the possibility of an invention being "difficult to imitate by its competitors." Such objectives have a lot in common with the field of economics, and include, but are not limited to, regional competitiveness, industrial competitiveness, corporate competitiveness, management competitiveness, service competitiveness, brand competitiveness, financial competitiveness, and quality competitiveness.

From the perspective of competitiveness, when designing or choosing the operating model, an individual enterprise should maintain its own competitive advantages, such as heterogeneity, uniqueness, and rarity, so as to obtain stable super-profits over a relatively long term.

8.1.3   Convergence

The term "convergence" is derived from biology. It means "the independent evolution of similar features in species of different lineages." Pitirim Sorokin, a Russian-American sociologist who founded the Department of Sociology at Harvard, introduced this concept into the sociology in his 1944 book *Russia and the United States* to elaborate on the similarity in the evolution of the two social systems as capitalism and socialism. The idea of convergence in economics thus came to acquire the meaning "to catch up," namely by "a reduction in the dispersion (or disparity) across economies."[6] Jan Tinbergen and John Galbraith, the former a winner of the first Nobel Memorial Prize in Economic Sciences (1969) and noted for launching econometrics, were the main supporters of convergence theory. To some extent, since all human activities increasingly "go digital" in the broadest sense of the term, such a reduction is being seen to occur in the unprecedented growth opportunity for the individual enterprise in the data industry, but also by the growing threat from other enterprises that can absorb another's core business to fill their own. Yet, and far more fundamentally, enterprises from distinct industries are also converging as boundaries disappear between philosophies of business and management styles, and all face similar force fields.

From the perspective of convergence, in designing or choosing an operating model, an individual enterprise should have a thorough ecosystem strategy and the right partners. Together they should be able to conform to the general trend of industrial development and any number of economic factors, (e.g., market system, industry development levels, and changes in supply and demand to eschew homogenization and low profits.

---

[5]http://en.wikipedia.org/wiki/Core_competency.
[6]http://en.wikipedia.org/wiki/Convergence_%28economics%29.

8.2   DATA INDUSTRY OPERATING MODELS

Choosing to enter the blue ocean is a natural selection of economies. According to W. Chan Kim and Renée Mauborgne [67] of INSEAD (European Institute of Business Administration), "red oceans represent all the industries in existence today – the known market space; and blue oceans denote all the industries not in existence today – the unknown market space, untainted by competition." Therefore many enterprises (e.g., IBM) are changing their operating models in order to reshape their business boundaries and conform to "the rules" of the future big data market. In this section we discuss some such rules to provide a richer picture of data industry operating models.

8.2.1   Gradual Development: Google

Most Internet data resources owned by Alphabet Inc. (Google's new parent company formed in August 2015 and thus considered the most valuable company worldwide in 2015) are beginning to collect data from Google Search.[7] Google Search is a raw data product that handles more than 3 billion searches queries each day (2012) – the figure still holds today, since "over" leaves things open ended (and by implication, higher).[8] In September 1998 Larry Page and Sergey Brin founded Google by relying on a clumsy algorithm called PageRank. PageRank[9] is "a link analysis algorithm that assigns a numerical weighting to each element of a hyperlinked set of documents." It is based on a hypothesis that higher quality pages get users' attention and thus get visited more frequently. Information related to Google's operating model like "Don't Be Evil," "Good enough is good enough," and "Google has no strategic plan,[10]" was revealed in Douglas Edwards's 2012 book *I'm Feeling Lucky: The Confessions of Google Employee Number 59*. Google told its investors that the only competitive objective with "no ideas disclosed" was to occupy over 50% of the market share in the next two years in searches "without doing evil." Such a unique operating model attracted several well-informed venture capital funds like Sequoia Capital and also Kleiner Perkins Caufield & Byers.

Google, from then on, gradually developed. With its diversification strategy, Google is stretching the full-field of all human activities using data technologies, such as adjusting search algorithms, hoarding data resources, and recruiting top-level data scientists. With regard to algorithm adjustment, Google acquired several companies like DejaNews, Blogger, Picasa, and Keyhole to optimize search algorithms from passive search to active learning. In order to hoard data resources, Google has increased the quantity and capacity of its data centers. Though Google does not publicly disclose the number of servers it has, in August 2011, watchers' guesstimates "placed the Google's server count at more than 1 million"[11] around

[7]http://en.wikipedia.org/wiki/Google_Search.
[8]http://searchengineland.com/google-1-trillion-searches-per-year-212940.
[9]http://en.wikipedia.org/wiki/PageRank.
[10]Word cited from Cindy McCaffrey, board member and employee number 26 of Google, lies in Chapter 4 of Edwards's book.
[11]http://www.datacenterknowledge.com/archives/2011/08/01/report-google-uses-about-900000-servers.

the world. In staff recruitment, according to a statistic at Statist.com,[12] as of 2014, Google has 53,600 full-time employees, which is more than triple the count in 2007.

At present, such products like Google Maps, Google News, and Google Translate have received critical acclaim. Google Maps[13] boasts several big improvements since its launch in 2005, including the incorporation of satellite imagery, street maps, a "Street View" perspective, and a sophisticated route planning tool for travel via foot, car, and public transportation. Most popular among them are the public transportation maps, which are especially useful in that roughly 500 cities (e.g., New York, Washington, Tokyo, Sydney) are covered, and they include arrival times for a combined 1 million or so stops for buses, trains, subways, and trams. Google News, launched in 2012, is completely computer generated. It collects and display headlines from thousands of news sources around the world, without manual intervention. Its metrics include[14] the number of articles produced by a news organization during a given time period, the importance of coverage from the news source, the average length of an article from a news source, breaking news score, usage patterns, human opinion, circulation statistics, and the size of the staff associated with a particular news operation. Google Translate is an automatic machine translation service that supports more than 90 languages. It utilizes a huge training data set and finds the most possible, but not necessarily the most accurate, translations in all circumstances.

The main problem of Google as a typical data enterprise is the overall size and complexity of the company. It is gratifying to know that the founders have discovered this deficiency and have proceeded to fix it by a restructuring starting in late 2015.

8.2.2 Micro-Innovation: Baidu

There is a popular saying among Chinese netizens: "if you have a question about internal affairs, ask Baidu; if you have a question about external affairs, ask Google."[15] The Chinese search engine accounted for 78.6% of the Chinese search market in 2012, compared to Google's only 15.6% share. Despite Baidu's high ranking bid, search methodology (i.e., all paid search results are processed ahead of free search results) has long been controversial, and it ranks merely as a "micro-innovation." The micro-innovation is not about revolutionary business models or technologies, it is rather a convergence in operating models. In other words, it combines imported ideas and local features to keep Chinese enterprises in step with the Silicon Valley tycoons, and perhaps someday it will secure a competitive spot in the industry.

Baidu's micro-innovation operating model, from the perspective of competitiveness, has a certain degree of heterogeneity that is reflected in (1) its search engine commercialization and (2) identity scheme. Baidu's slogan of "Behind your e-success" was established to serve enterprises rather than the general public. Among

---

[12]http://www.statista.com/statistics/273744/number-of-full-time-google-employees.

[13]http://en.wikipedia.org/wiki/Google_Maps.

[14]http://www.computerworld.com/article/2495365/business-intelligence/an-inside-look-at-google-s-news-ranking-algorithm.html.

[15]http://theconversation.com/baidu-eye-micro-innovation-or-copying-google-glass-13303.

these are Sina, Sohu, NetEase, Tom, and Yahoo! China, and other institutions include CCTV (China Central Television). In use of language, Baidu closely cooperates with Chinese authorities, usually *People's Daily*, and thus is able to master Chinese keywords better than Google. The tastes and preferences of China's market also tend to favor Baidu (e.g., Baidu Video Search, Baidu Tieba, Baidu Baika) to Google. This is because, Chinese users mainly use the search engine for entertainment, instant communication, and "tucao" (meaning "ridicule" in Chinese) and so are accustomed to using free services.

8.2.3   Outsourcing: EMC

The EMC Corporation, founded in 1979 and headquartered in Hopkinton, Massachusetts, has been working on outsourcing as its operating model to face the challenge of big data for about several years now.

Since 2007, EMC has been outsourcing its famous Digital Universe project[16] – "the only study to quantify and forecast the amount of data produced annually" – to IDC, an American market research firm providing analysis and advisory services. In 2013, Chenghui Ye, EMC's global senior vice president and president of the Greater China Region, announced "a new goal for China."[17] The goal is to expand the local base to be the biggest base of EMC outside the United States, and to increase channel partners that participate in the distribution of EMC storage products from the current 3,578 to about 20,000 in the next five years. This goal, in point of fact, conforms with China's emerging markets; China accounted for 23% of the "digital universe" in 2010, expanded to 36% in 2012, and will reach 63% by 2020.

What EMC elected to do is consistent with EMC's original strategy to expand its market share using competitiveness and convergence. However, EMC did not add any new business module. "Just because a market is big and growing; it does not mean it has profit potential." EMC's focus on commodity hardware destines it to make very little profit. Unfortunately, the US$67 billion acquisition of EMC[18] that Dell proposed in October 2015 did succeed.

8.2.4   Data-Driven Restructuring: IBM

Facing big data challenges, the International Business Machines Corporation (IBM) holds a diametrically opposite viewpoint relative to EMC. IBM has adapted thorough business adjustment strategies, namely by data-driven restructuring in all of its businesses.

Since the transnational firm was founded in 1911, it has grown to over 430 thousand employees in 175 countries. IBM "reduced its global workforce in 2014 by more than 12 percent, according to a recent filing with the US Securities and Exchange

---

[16]http://www.emc.com/leadership/digital-universe/index.htm.
[17]http://www.chinatechnews.com/2013/03/06/19165-emc-chinas-growth-focuses-on-big-data-cloud-computing.
[18]http://www.forbes.com/sites/petercohan/2015/10/18/why-the-dell-emc-deal-is-doomed-from-the-start/#14433fc031b5.

Commission."[19] It completed more than thirty major gradual transactions[20] from 2006. These include FileNet in 2006 (content management and business processes), Cognos in 2007 (business intelligence and performance management), SPSS in 2009 (statistical analysis), Coremetrics in 2010 (web analytics and marketing optimization), Netezza in 2010 (data warehouse appliances and analytics), OpenPages in 2010 (governance, risk management, and compliance), i2 Limited in 2011 (visual intelligence and investigative analysis), Algorithmics in 2011 (risk management), DemandTec in 2011 (lifecycle price management and optimization), Emptoris in 2012 (strategic supply, category spend, and contract management), Varicent in 2012 (incentive compensation and sales performance management), Vivisimo in 2012 (search engine), Tealeaf in 2012 (customer experience management), Butterfly Software in 2012 (data analysis and migration), StoredIQ in 2013 (records management, electronic discovery, compliance, storage optimization, and data migration initiatives), Star Analytics in 2013 (process automation and application integration), Softlayer in 2013 (cloud infrastructure), CrossIdeas in 2014 (identity governance and analytics), Silverpop in 2014 (email marketing and marketing automation), and Explorys in 2015 (healthcare intelligence), Merge Healthcare in 2015 (medical image), and Weather.com in 2015 (weather data). IBM agreed to sell its x86 server business to Lenovo for US$2.3 billion in January 2014. In October 2014, IBM paid GlobalFoundries, a semiconductor foundry, US$1.5 billion over the next three years to dump its chipmaking business. Additionally, it is rumored that IBM will give up its consulting business in the near future.

IBM[21] is eyeing investing US$40 billion in the Cloud, big data, and security revenues by 2018; it has already benefited from such data-driven restructuring, and even attracted an investment by Warren Buffett, who had never before purchased tech shares.[22] While other companies were merely looking to data as the new means of competitive advantage, IBM had already found a growth engine that drives faster innovation and in wide applications. These growth points include, but are not limited to, determining retail store locations, probing hot crimes, predicting student achievement, and combing hospitals for infection sites.

8.2.5   Mergers and Acquisitions: Yahoo!

Mergers and acquisitions (M&A) have in the past served as quite effective strategies. In times of hot competition from other enterprises for market share, M&A can help an enterprise grow rapidly in its sector at its original location, or in a new sector and location without a lot of capital investment, human, financial, and otherwise.

Yahoo!, established in 1994, was a miracle maker of the Internet at the end of the twentieth century. Yahoo! was a strong supporter of M&A, but its M&A activities

---

[19]Reduced its global workforce in 2014 by more than 12 percent.
[20]An organic transaction occurs when an individual enterprise acquires another one that has been integrated into others businesses in any sort of way.
[21]http://www.firstpost.com/business/ibm-eyes-40-billion-cloud-big-data-security-revenues-2018-2130847.html.
[22]Berkshire Hathaway owns 76.9 million IBM shares as of the end of 2014, for a market value of more than $12.3 billion and a company ownership of 7.8%.

were not always smooth sailing. Yahoo! experienced many failures from its M&A activities of which the five biggest[23] were Delicious (bought for a rumored US$30 million in 2005 and sold for a questionable UK£1 million in 2011), Flickr (bought for a questionable US$35 million in 2005), MyBloglog (bought for US$10 million in 2007 and closed in May 2011), Kelkoo (acquired for US$672 million in 2007 and sold for "less than US$100 million" in 2008), and Broadcast.com (acquired for a reported US$5.7 billion in 1999, one of the biggest deals in the history of the Internet), and other troubling cases included GeoCities (currently available only in Japan), Dialpad (a discontinued Yahoo! service), and 3721 Internet Assistant (malware). Yahoo! once attempted to acquire Google, Baidu, eBay, and Facebook but was rebuffed, and later thwarted by these companies. The most successful M&A case of Yahoo! was the Alibaba Group, on which it spent US$1 billion in acquiring a 40% value of equity securities in 2005, and obtained dozens of returns amounting to US$7.6 billion in repurchasing 20% equity in 2012.

Despite having a high probability of failure, it should be noted that Yahoo!'s use of M&A is still a good operating model in that it has increased profits around its own core strategies. This is reflected in, first, Yahoo!'s recruiting talents and improvements to technologies, significantly upgrading their innovation capability; second, Yahoo!'s acquiring rapidly growing enterprises to evade subversive competitors; third, Yahoo!'s using similar size M&A to form a complementary enterprise or to reduce the intensity of competition in its field. It should be noted that the use of M&A for the sole purpose of large companies as an alternatives of investing in R&D themselves, is not desirable.

8.2.6   Reengineering: Facebook

Facebook has over 1 billion worldwide members, masks off Google, and has become an independent data kingdom. All businesses of Facebook are deemed to be data developments. Everything has its pros and cons. Massive amounts of high-quality social network data that Google has been craving are accompanied by two disadvantages: (1) the difficulty of user conversion, and (2) data privacy. User conversion refers to the conversion of site visitors into paying customers. The difficulty is reflected in the fact that Facebook visitors have generally lower purchasing needs than consumers of search engines. This is a reason why Facebook's advertising can never become the major source of its operating income. In regard to data privacy, the core value of Facebook's social network data itself is truthfulness. Private information contained in Facebook includes gender, age, personal experiences, interests or preferences, and friends.

In May 2007, Facebook opened its platform of 130 million active users, and provided third-party developers access to core functions through an API (application programming interface) for application development. Some people thought that Facebook was going to follow Microsoft by this act, and claimed that its aim was to be "another Windows." Others believed that Facebook's platform was to be an "Apple's

---

[23]https://econsultancy.com/blog/8811-yahoo-s-five-biggest-acquisition-screw-ups.

iOS" imitation. Surprisingly, Facebook actually regarded Google to be a competitor. In January 2013, Graph Search, a new data product, launched by Facebook with a lofty tone, coupled with the earlier product "Timeline." The two products significantly enhanced users' viscosity of Facebook.

As is clearly shown, Facebook essentially adopted an operating model of reengineering its competitors' core businesses. The reason for not describing it as "business copying" is that the data resources of Facebook are totally different from its competitors. Therefore, after an ugly 15 months of trading,[24] the Facebook stock price not only passed its US$38 initial public offering price but more than doubled in 2013 and its future prospects remain promising.

8.2.7 The Second Venture: Alibaba

The Alibaba Group began in 1999 and it primarily operates in the People's Republic of China (PRC). On the date of its historic initial public offering – September 19, 2014 – at closing time, its market value was measured as US$231 billion.[25] However, before the public listing Alibaba's founder and CEO Jack Ma retired, as was announced on May 10, 2013. Nevertheless, eighteen days after stepping down as CEO, Jack Ma come back and announced a plan for Alibaba's second venture.

Alibaba's second venture was actually planned at the beginning of 2012. In February 2012, Alibaba announced a privatization offer in repurchasing equity at an average cost of HK$13.5 per share. On May 21st, Alibaba entered into the final agreement with Yahoo! to repurchase a 20% value of equity in the amount of US$7.6 billion. On July 23rd, Alibaba announced its reorganization plans under the designation CBBS, which stands for "consumer to business to business to service partners," in order to adjust its seven business groups, including Taobao.com (C2C), Etao.com (shopping search engine), Tmall.com (B2C), Juhuasuan (group-buying), Alibaba International Operation (global platform for small businesses), Alibaba Small Business Operation (e-commerce service for domestic SMEs), and Aliyun (cloud computing). Then in April 2013, Alibaba purchased an 18% stake in Sina Weibo for US$586 million, as it moved to broaden its mobile offerings. On May 11th, the Alibaba Group agreed to acquire AutoNavi in a deal that is about US$294 million, with a further issue of preferred shares and ordinary shares accounting for 28% after expansion, to bolster Alibaba's Internet mapping tools. On May 28th, Alibaba in tandem with industry partners announced the establishment of a 100 RMB billion "smart" logistics network that aimed to make 24-hour domestic deliveries possible.

Jack Ma's smoke and mirrors game was finally settled. He had essentially built a huge "data" empire that could implement data innovations for trade, finance, and logistics. In financial services alone, according to the data provided by the Alibaba Group, at the end of 2012, (1) Alibaba China had 5.2 million registered members, of which 8 million were corporate members and 650 thousand were TrustPass members; (2) Alipay had over 800 million registered accounts, of which the peak transaction

---

[24]Shares of the social networking giant have done a stark about-face since hitting a low of $17.55 in September 2012.
[25]http://en.wikipedia.org/wiki/Alibaba_Group.

volume reached 105.8 million per day; (3) Tmall.com had exceeded 60 thousand enterprises; and (4) Taobao.com had exceeded 7 million sellers. Alibaba Group's financial affiliate was built on the huge users and transactions, which are invaluable data assets and difficult to be copied by other competitors. Meanwhile, Alibaba started rebuilding a credit system using e-commerce data and honesty records to implement "data loans," contrary to the traditional loans heavily used by securities or mortgage banks. Alibaba recently released its data capabilities for the public using a service platform called DataPlus (2016) that aims to minister to China's big data market of nearly 1,000 billion RMB over the next to or three years.

In brief, Alibaba successfully transformed into a typical enterprise within the data industry in using data innovations to reshape industry boundaries for enhancing its value creation level.

8.3   INNOVATION OF DATA INDUSTRY OPERATING MODELS

In a free market, enterprises must keep in step with a changing business environment due to such matters as upgrading to meet consumer demand or preferences, exacerbating interactions with competitors, invigorating the performances of technologies or the levels of industry, tweaking internal policies to shorten the product lifecycle or to attenuate the target market, and so on and on. In this sense, enterprises must continually update their operating models.

As was noted earlier in this chapter, the operating model of enterprises in different industries is quite similar in that ever so slight differences pertain to the unique features of a domain or sector. These include changing original strategies, adjusting existing resources, refreshing obsolete technologies, and rebuilding competitiveness.

The innovative uniqueness of data industry's operating models is reflected in data application innovations, the changing philosophy of business, management styles, and the force fields of enterprises in related areas or industries and even in traditional industries.

8.3.1   Philosophy of Business

A philosophy of business is fundamental in that it is the principle motivation for the formation and operation of the business enterprise. It expresses the nature and purpose of the business, and the moral obligations that pertain to it.[26] However, it is irrelevant that the organization aims to reap profits.

In the data enterprises, the philosophy of business has to further include scientific prediction and mission recognition. A prediction is a statement about a future event. In this regard the enterprise has to find the point of data that through innovative scientific intervention it can reliably design a specific data technology roadmap to optimize technologies, products, markets, and their interactions of the original industry, via existing resources. There are two things to remember in preparing a mission statement. First, enterprises in traditional industries should identify any impediments

---

[26]http://en.wikipedia.org/wiki/Philosophy_of_business.

to their development and their own weaknesses, and propose timely transformation actions and upgrades using data technologies. Second, emerging data enterprises cannot focus on short bursts of profit growth but on stabilizing long-term development.

8.3.2 Management Styles

Management styles are characteristic ways of making decisions and relating to subordinates. Management styles vary across organizations and even stages of organizational development. They can be categorized into many personal styles, such as autocratic, paternalistic, and democratic.

In data enterprises, management styles should be permissive and thus cultivate a relaxing work environment with flextime as needed. Business executives themselves may adopt data innovations in routine work such as (1) in the recruitment and selection of employees, by applying text analytics to filter resumes, search engine to double-check candidates, and association analysis to match job vacancies and STEM skills,[27] and then (2) in team building, by utilizing mining algorithms to subdivide preferences and capacities of employees and to carry out best resource allocation on schedules, skills, and wages.

8.3.3 Force Field Analysis

Force field analysis is a useful management tool that can be used to find appropriate driving forces and then to analyze the pressures for or against an organization. These driving forces are usually inevitable, and they may include natural, physical, and even biological driving forces.

In data enterprises, force field analysis can galvanize data scientists to find active solutions and thus serve to stimulate creativity. Force field analyses can also address issues of human resource management by using data innovations in explaining employee estrangement (e.g., due to routine work, a lack of input in basic decision making, or manager–employee friction), in attending to Maslow's hierarchy of needs, or even in initiating a multi-factorial incentive system.

---

[27]STEM refers to science, technology, engineering, and mathematics.

9

ENTERPRISE AGGLOMERATION OF THE DATA INDUSTRY

Industry clusters, unlike the industrial chain, depend less on industry interrelatedness. They consist of a collection of both connected and unconnected industries located geographically together as a community of industries. Four relevant models of such industry communities are discussed in this chapter: agglomeration, enterprise cluster, industrial concentration, and industry aggregation.

Alfred Weber, a German economist, formulated a set of rules and concepts relating to agglomeration in his 1929 book *Theory of the Location of Industries*, which he defined as the process of spatial concentration. Enterprise cluster theory treats a single enterprise as an independent life, so "interactions between different enterprises as well as the environment" are analogous to the phenomenon of aggregation of species in nature. The distribution of such enterprise clusters can involve different industries, however. Industrial concentration refers to the situation where several relatively large-scale enterprises in an industry locate in close proximity to each other. Industry aggregation research is a study of the close spatial distribution of same or similar industries, and the research especially considers the degree of concentration changes: from dispersed to concentrated.

Industrial concentration that results in an industry monopoly is irrelevant for spatial distribution studies. Industry aggregation may not always bean industry cluster because, to form a cluster, "the interactions between different enterprises as well as the environment" must be connected. So agglomeration is just the starting point of industry aggregation research. The same is true of the data industry.

The Data Industry: The Business and Economics of Information and Big Data, First Edition. Chunlei Tang.
© 2016 John Wiley & Sons, Inc. Published 2016 by John Wiley & Sons, Inc.

Figure 9.1   Directive agglomeration. The dashed outline shows a location area with certain resources, the × symbol shows the product-specific interactions, and the triangles the locations of the enterprises

9.1   DIRECTIVE AGGLOMERATION

In his 1990 book *The Competitive Advantage of Nations* (1990) [68], Michael Porter gave a case study of the Italian tile industry that has been widely cited. These Italian tile producers were mainly concentrated in the Sassuolo district, Emilia-Romagna. Materials and tools are heavily dependent on imports, including the kaolin clays (a raw material for tile), furnace, flattening machine, and even glazing machine. However, thanks to Italy's rebuilding plan after the Second World War, the domestic demand increased sharply. A major force behind the spontaneous growth of the market was the geographical concentration in Sassuolo. There the tile industry gained and maintained a strong competitive edge because of not only the were located supporting sectors offering molds, glazes, packaging materials, and transportation services but also some small professional firms for management consulting, outsourcing, and other professional services. Such a typical directive agglomeration is shown in Figure 9.1. In the data industry, there are two unique characteristics in directive agglomeration: data resource endowment and multiple target sites.

9.1.1   Data Resource Endowment

Resource endowment is used to identify in a location area whether the factors of production are abundant or poor. A factor of production is considered abundant if a relatively large proportion of the supply is at a lower price than in other locations; otherwise, it is considered to be poor. In factor endowment theory, as originally developed by Swedish economist Eli Heckscher and his student Bertil Ohlin, resource endowments that general industries rely on include (1) human resources at a low cost as specialist labor, (2) physical infrastructure at a proximate origin as raw materials, (3) access to transportation to market concentration areas or transportation hubs, and (4) technological surroundings for spillover and demonstration effects.

Compared with traditional industries, the data industry is concerned with data resource development and data asset management. That data must rely on its carriers for storage, a location area (i.e., a region) with big data centers, (i.e., whereas other locations might not), and procession of an additional unique data resource endowment (i.e., data resources and data assets). Of course, there is the prerequisite that there should already be a data resource endowment.

9.1.2 Multiple Target Sites

The term "target site" is used to describe "a specific need," and the need cannot be limiting but be diverse and varying among customers. A major difference between the data industry and traditional industries is that the target is not just a product-specific need but simultaneously meets various customer demands based on multiple levels of understanding as to what is a specific need.

"Education may affect the basic demand for health" [69] is a specific need, for example. Different enterprises of the data industry will attempt to target several levels of understanding, such as (1) some enterprises that own health data resources may promote distribution of accurate health information to well-educated people; (2) some enterprises involved in education may offer MOOCs (massive open online course) to the poorly served parts of the population by higher educational facilities; (3) some enterprises that control both health and educational data resources may elect to exclusively mine the relatively more diverse interests of well-educated consumers who want to learn more about health or about management consulting. Thus, this type of enterprise agglomeration in the data industry will cluster various enterprises that provide diverse data products on a specific need with multiple target sites.

9.2 DRIVEN AGGLOMERATION

Driven agglomeration follows from favorable exogenous conditions inducing a geographical concentration of enterprises. The typical examples include the Shipbuilding SMEs of Croatia affected by government policy guidance or spinoff deployment in administration, and the Hong Kong financial services cluster arising from international investments. These propitious conditions vary and can be a dis-/or agglomeration factor, depending on whether a centrifugal tendency is opposed by a centripetal force [70]. Generally, agglomeration is fostered by labor proved to be capable, labor specialization, technology spillover, identical levels of culture, and innovative practices already in place. Dispersions of enterprises are often due to population crowding, rising land prices, market restrictions, technology interruptions, overcompetition, and resource constraints.

The data industry will likely agglomerate its billions of small-but-excellent and fast-and-accurate enterprises in the near future. The studies continue in search answers to the problem of accelerating development of such micro, small, and medium size enterprises at geographic locations with favorable exogenous conditions, and at the same time avoiding friction and collisions due to shared market appeal. At the present time, research is needed on the three main forces comprising human capital, financial resources, and the scientific and technological developments, so as to create synergy.

9.2.1 Labor Force

Human capital, namely labor force, is a resource of social wealth that depends on the creativity of the human being. A creative labor force is important for any industry, and without exception for the data industry. Any enterprise generally elects to settle in regions where the population is relatively well educated and capable of group learning. In this regard, the data industry is like the traditional software industry in terms of the composition of the labor force, labor costs, and competition in the labor market, as both have high-technological content and high permeability.

Software enterprises usually have three types of labor force: (1) personnel who understand both technology and management, (2) staffs who can design or analyze architecture systems with strong technical skills and a fluency in computer theory, and (3) blue-collar workers who are trained by vocational-technical schools. Data enterprises that aim to develop massive data with heterogeneous and complex structures additionally require data scientists. Data scientists need to face both customers and the data, and mainly handle work in two parts. The first is to discover appropriate data-processing methods based on the strength of their data-driven insights. The second is to design or optimize mining algorithms to find hidden regular patterns that support business applications. Data scientists typically have graduate academic degrees relating to data processing skills in computer science (especially in data mining), mathematics, physics, statistics, or life sciences, and additional skills in technology, business, and customer relations.

The development of software is usually organized by a meticulous clear division of labor. Particularly, for large system development, the whole team is required to work for several months or even years. Staff flow and the loss of key technical personnel are high risk. This means that labor costs are more important than materials for software enterprises. Data enterprises are generally biased in favor of applications in a wide variety of domains or sectors, and have the following characteristics in routine works. (1) Only a few algorithms are required to be developed by data scientists from scratch, since the technical boundary is greatly extended by open source tools. (2) Labor costs are reduced through data enterprise crowdsourcing, which is manifest by many amateur analysts serving as temporary staffs or partners. Therefore labor costs of data enterprises can be brought slightly lower.

Competition among software enterprises is quite intense, and this is often reflected in the competition for talent of senior architects and senior software engineers, who may be indispensable for a software development project. However, in the data industry, despite having the same urgency for top quality personnel, different data scientists may find different hidden patterns based on their own respective data-driven insights, so no one is "indispensable."

9.2.2 Capital

Financial resources in exogenous conditions are the capital investments. Here, we consider a generalized capital stocks concept that diversifies its investments including both physical and nonphysical assets. Correspondingly, what is obtained from investments is the return on capital (ROC). Generally, capital return is of risk, meaning investors may either obtain excess benefits from the time value of money or suffer loss.

From the perspective of external funding sources, capital includes debt and equity capital. Debt capital is mostly from financial institutions, and equity capital can be classified into public offering and private placement. For a startup enterprise, the general characteristic of high risk does not match the conservatism principle of prudence in banks, and its scale is difficult to meet the tougher reporting and governance standards of public listing. Therefore both debt capital and public offering of equity capital cannot become a start-up's main financing means. Start-ups usually depend on the venture financing systems composed of the 3Fs (founders, family, and friends), the government and government agencies, angel investors, and private equity strategic investments, in order to realize private placement of equity capital.

In view of investment and financing strategies of risky private placement, the investors choose respective investment approaches for different investment stages. This involves necessary management output according to the degree of risk in the course of the start-up's growth and development. Regulatory risk evaluation usually undergoes a relatively long demonstration process, such as meeting the founder's team, analyzing business plans, having a field visit, consulting third parties, and investigating market research.

The data industry start-ups usually seek a "simple and rapid" type of professionally managed industrial investment fund that can provide rapid evaluation of technical content and a clear investment direction at scale, that is more attractive than risky private placements in that the target selections are wide ranging, involve industry dispersion, and require repetitive demonstrations. Furthermore the first round of professional industrial funding has three advantages: (1) it is conducive to helping start-ups attract subsequent investment supports, (2) it is conducive to the establishment of a rational exit mechanism with a reasonable capital cycle, and (3) it is conducive to innovative business models or operating models, that promote industry to university collaborations and lead to achievements in data science and technology.

9.2.3 Technology

Intrinsic to the commercialization of data products have been the exogenous driving forces of the sciences and technology. Realistically, however, the commercialization of the data industry to serve economic and social advancement, as most people expect, is only a dream.

Small and medium-size enterprises (SMEs) in the current climate are hanging by a thread, and their CEOs tend to be unwilling to extract capital stock from the extremely weak financial chain to implement R&D input. In terms of Joseph Schumpeter's business cycle theory, enterprises need innovation, but it must be an "introduced type" of innovation. Initially, innovation has to be separate from profit-making activities. In other words, it is the enterprise that is responsible for the profit-making activities, and the universities and scientific research institutes to complete the former – the initial innovation. This is the only way that the initial technological innovation can become an external force that draws and binds an enterprise agglomeration.

Unlike other technologies, the innovation, conversion, and application of data technology, first, is to directly respond to application areas; there are no patents or drawings to file, technical know-how or demonstrations to document, testing of products for endorsement or management consent, but only directly marketable application results. Second, to directly respond to data resources, the selection of labor, materials, and capital, only interested persons or enterprises can participate in the public tender. Third, to directly respond to consumer demand and preferences, the firm that adopts fast technological innovation must have low-risk production and marketing aims.

9.3 INDUSTRIAL SYMBIOSIS

The word "symbiosis" was coined by the German mycologist Heinrich de Bary in 1879. Symbiosis refers to an interdependent system formed by "the living together of unlike organisms"[1] due to survival needs. Over the past century, theoretical research on "symbiosis" has penetrated and stretched from the field of biology to encompass the economy and society. It is used to probe occurrence regularity, development law, and human interaction with the environment in addressing issues of the symbiotic relationships between organizations and society as well as the issues of complementation and innovation.

Industrial symbiosis aims for coexistence, coevolution, and convergent evolution under a certain value chain. Industrial symbiosis is mainly affected by three factors: (1) degree of adaptability of the individual enterprise, (2) the market environment, and (3) competitive or cooperative interactions among enterprises. In accord with the mechanism of enterprise agglomeration, the formation of industrial symbiosis can be of two types: (1) as entity symbiosis, whereby the relationship may be familial, industrial, or geopolitical, or (2) as a virtual derivative, whereby the symbiosis develops from network connectivity.

9.3.1 Entity Symbiosis

Entity symbiosis is interest driven and cooperatively balances activities as they evolve and develop. A certain incidence relationship between enterprises is the original intention for building a symbiotic system. With the establishment of a symbiotic system, the individual enterprises are not self-dependent. It is a game of matching complementary resources for mutual benefit of member enterprises, bound by understanding each other's deep vision, mission, and values.

Entity symbiosis commonly has three relationship forms: familial symbiosis, industrial symbiosis, and geopolitical symbiosis. There is a familial relationship that binds member enterprises in the familial symbiosis, such as affiliate, associate, and subsidiary enterprises. Familial symbiosis can make it easier for participants to accept each other's cost-cutting practices and to share benefits. Industrial symbiosis usually consists of member enterprises having business dealings upstream and/or

---

[1] http://en.wikipedia.org/wiki/Symbiosis.

downstream in the same industrial chain. It aims at various strategies to reduce market risk and improve mutual economic benefits. Geographical symbiosis is composed of member enterprises in a certain region that share public resources and an environment for diversified cooperation, and usually results from the demand of a single major enterprise.

The formation of an entity symbiosis system by a combination of traditional industries with the data industry is an optimum from the perspective of the strategic emerging industries. This is because, by actively cultivating enterprises in traditional industries, it may be possible the data industry start-ups to upgrade and transform the core technologies for specific, as yet unmet applications. Future data enterprises will be flexible, adaptable, and innovative, and can solve the issues of enterprises in traditional industries (e.g., supermarkets, taxi companies, iron and steel enterprises) based on diversified cooperation, and extending to internal reform, operation management upgrades, expanding varieties, increasing quality, and reducing the consumption of raw materials and energy usage.

In brief, establishing a strategic alliance based on entity symbiosis, and led by leading enterprises of both the data industry and traditional industries, may turn out to be better for the economy than a pure data industrial association "without cooperation but competition."

9.3.2 Virtual Derivative

The virtual derivative type of symbiosis goes beyond all familial, industrial, and geopolitical relationships to form a loosely coupled system established through a network selection mechanism. The virtual derivative corresponds closely with the market demand of an enterprise. Its essence is that a single enterprise depends on other enterprises with related interests in order to optimize and integrate external resources, and to remain flexible enough to keep growing.

As noted by John Byrne [71] in 1993, the former executive editor-in-chief of *Business Week*, "a virtual enterprise is a temporary network of independent companies" for the purpose of seeking maximum adaptability. Member enterprises, the composition of a virtual enterprise, are responsible for mutually independent business processes using their respective "core abilities," to realize skill-share and cost sharing. The virtual enterprise commonly has the following two characteristics. First, it breaks up the tangible boundary of enterprises, in that a virtual enterprise has neither an organizational boundary nor an organizational structure. Second, the temporary combination can avoid "big company disease" as defined by Parkinson's law (on the rate at which bureaucracies expand over time by officials who multiply subordinates and make work for each other).

In the data industry, this type of enterprise agglomeration needs data innovations that obtain individualized and diversified demands of various environmental elements. These include choosing appropriate partners, relying on institutions of knowledge production, obtaining new technologies and complementary assets, dispersion of innovation risk, overcoming or building market barriers, forming smart manufacturing, and launching new data products at a low cost but with high quality.

9.4 WHEEL-AXLE TYPE AGGLOMERATION

A wheel-axle type agglomeration looks like the wheel and axle of an automobile in terms of structure. This layout is conducive to the practice of the input and output operations that describe the upstream and downstream enterprises. Generally, the upstream critical enterprises are the headquarters.

9.4.1 Vertical Leadership Development

Detroit, MI, and Pittsburgh, PA, are both typical wheel-axle type agglomerations. They are characterized by (1) an obvious hierarchy between critical enterprises and other enterprises and (2) several critical enterprises in a strong position to vertically lead and decide the development of the whole city.

The Detroit area, strategically located along the Great Lakes waterway, emerged as a significant metropolitan region within the United States in the early twentieth century, and this trend hastened in the 1950s and 1960s when the Big Three automobile enterprises of General Motors, Ford, and Chrysler, one after another, established their headquarters in Detroit. The city of Pittsburgh, surrounded by three rivers – the Allegheny, Monongahela, and Ohio – is known as "the Steel City" for its formerly more than 300 steel-related businesses, including the large heavy industry consisting of US Steel Corp., Westinghouse, and Alcoa. In the early 1970s, the two cities began differentiating: Detroit continued developing the automobile industry; while Pittsburgh diversified and embarked on an urban "renaissance," led by Carnegie Mellon University and the University of Pittsburgh, with regional policies formulated toward developing high-tech industries like cyber defense, software engineering, robotics, energy research, and biomedicine. Bankruptcy in Detroit resulted in the closure of factories and the disappearance of jobs, whereas Pittsburgh economically rebounded and has been named one of "America's Most Livable Cities"[2] six times since 2000 by *The Economist*, *Forbes*, and *Places Rated Almanac*. Why? Detroit had unfortunately "put its fate in a few critical enterprises," and by the auto industries' vertical "long-arm" leadership, other enterprises only passively absorbed a bit of the exogenous technical forces brought in by the "long arm." Pittsburgh gave up such a "long arm" and induced an outer force of the SMEs to participate in its transformation, and thus regained a competitive edge.

The wheel-axle type agglomeration is used in the data industry. However, the "wheel" and "axle" should be redefined so as to upgrade and transform the critical enterprises in traditional industries in line with the micro and SMEs mode of operations.

9.4.2 The Radiation Effect of Growth Poles

The growth pole theory, contributed by the French economist François Perroux, is that "economic development, or growth, is not uniform over an entire region but rather

---

[2]http://www.nextpittsburgh.com/business-tech-news/economist-names-pittsburgh-livable-city.

takes place around a specific pole (or cluster)."[3] This growth pole plays a dominant role in generating a force field within a single economic locale.

Strengthening the radiating effects of numerous growth poles may be the best way of developing regions, where the government is lax. Today, in China, governments at all levels have been taking charge in situating growth poles in order to invigorate the economy and obtain the economic benefits that go along with such strengthened industrial centers. However, some of China's second- and third-tier cities are more interested in introducing the established industry headquarters rather than cultivating new industry. The problem this introduces is twofold. First is the mobility of a headquarters introduced this year that may be stolen by another city the next year. Second is that where a zero growth economy already exists at the macroscopic level, the economic aggregate will not change (even loss) despite the headquarters relocation there. However, with the radiation effect of growth poles, there could be interregional cooperation among local governments beyond geopolitics, in place of just an industry's relocation. In other words, a growth pole for the data industry could drive a technological network in the shape of transfer, absorption, integration, coordination, and feedback interactions.

The growth pole's radiation of the data industry is therefore essentially a cross-region replication matter. However, cross-region replication is not just about making an exact copy, but replication with local characteristics. Take the taxi industry as an example. We assume that a taxi company has data innovative capabilities by understanding resident travel within a city using a historical trajectory of data from routine taxi operations. Consider replicating this data technology to another region; the radiation effect then could be manifest in the following steps: (1) introducing a case study, (2) analyzing local taxi trajectory data, (3) optimizing algorithm according to the data, and (4) finding the traffic jam hotspots, and local driver and passenger preferences. This same procedure could be applied in the practice of cultivating local taxi enterprises.

9.5 REFOCUSING AGGLOMERATION

Refocusing, a strategy due to British scholar Constantinos Markides, is not merely anti-diversification but an amendment to over-diversification with the aim of improving overall competitiveness, despite being the opposite of diversification. Though Markides's description is at a corporate level [72], applying to "the strategy for a company and all of its business units as a whole," in regard to enterprise agglomeration, refocusing is also a useful means of fixing the agglomeration problem where there is no industry but only separate enterprises.

9.5.1 "Smart Heart" of the Central Business District

The modern central business district (CBD) is often deemed as a city logo, symbol, or sign in the era of globalization, including Lower Manhattan in New York, La Defense

---

[3]http://people.hofstra.edu/geotrans/eng/ch2en/conc2en/growthpoles.html.

in Paris, Shinjuku in Tokyo, and the Central District in Hong Kong. The cohesiveness of a CBD as a physical-spatial form (e.g., location, layout) or social-spatial structure (e.g., natural environment, historical background, economic foundation, social culture) has long been recognized.

Peter Hall, a British scholar praised as a "world-class master of urban planning," once pointed out that a CBD should be multi-functional, ecological, and humanized. In other words, it should be a commercial district as well as a mixed-use district that permits residential, cultural, institutional, and industrial uses – not only hold various business activities but additionally provide entertainment, shopping, bodybuilding facilities, all in all, be a humanized community with strong cultural atmosphere. There is further another kind of consensus that a CBD should be equipped with the latest technology, and offer internal services of property management, community security, and resident demand in a "smart" manner – be a "smart" CBD. The CBDs can become "smart," but differently than many people have expected with the addition of a smart heart – the data technology, apart from the hardware.

9.5.2 The Core Objective "Besiege"

In recent years, the Oracle Corp and SAP AG have been worrying many smaller companies. Among these small companies are Workday, Inc. and Salesfore.com. Oracle specializes in developing and marketing its own brands of database management systems. SAP is a famous ERP (enterprise resource planning) solution provider that helps enterprises manage business operations and customer relations. Workday, founded in 2005 by co-CEOs Aneel Bhusri and Dave Duffield, offers workforce costs and manage staff pay analyses. Salesforce.com is a cloud computing company. In September 2013, Workday and Salesforce.com unveiled a unique partnership co-engineering an integrated SaaS platform with a "complete, end-to-end, enterprise cloud delivery." At the same time, both Oracle and SAP made a number of SaaS acquisitions. This situation shocked many industry people because they were more used to seeing large enterprises scuffle. Was there mismanagement at both Oracle and SAP? The answer is no, as both are rich and powerful. In fact in this 2013 case, a quarter-of-a-trillion dollars a year were plowed into integrating and managing Oracle and SAP.[4] Oracle & SAP vs. Workday & Salesfore.com respectively belong to reverse industries: the information industry and the data industry. The information product is Oracle's database system that is in common use and expensive, whereas the data product is highly targeted and relatively cheap pricewise (e.g., Workday makes money by offering subscriptions to services rather than selling software). Information companies always like to play a lone hand, whereas data enterprises often besiege a subset of main enterprises. Oracle and SAP were, however, being besieged by a couple relatively powerless the SMEs – and Oracle's several coping strategies almost all failed, key among these being reducing the database support services and implementing a hostile takeover of Workday. This is because a product that has a little defect but is practical and cheap will still be considered a good product.

---

[4]http://www.horsesforsources.com/salesforce-workday-hookup_092313.

Refocusing the core objectives in the data industry can bolster enterprise agglomerations and thus stimulate economic growth in their locales. The core objectives to consider in setting up an industry cluster are the site's data resources, the number of enterprises there already connected to traditional industries, and where the potential customers are located.

10

CLUSTER EFFECTS OF THE DATA INDUSTRY

An industry cluster[1] is more than a simple aggregation of enterprises. It is a geographic concentration of interconnected suppliers, providers, and associated institutions in a particular field through a nested inter-organizational network of relationships. As such, it represents a "hybrid" economic sphere, somewhere between a marketplace and an organizational hierarchy.

Once an industry cluster is formed, resource sharing or technology diffusion will follow based on the internal division of labor, competition, and cooperation. The clustering of interconnected enterprises will generate revenues through lower transaction costs, thus create huge external competitive advantages for these enterprises and likely for regional economy and even national politics, society, and culture. Therefore the core task of an industry developing a cluster is to study the positive and negative regional effects as well as larger external institutional factors.

Although at the present time there is no known cluster in the data industry, based on the enterprise agglomerations types described in the previous chapter, we can infer what conditions might facilitate the formation of data industry clusters, and avoid blindly copying other clusters of unrelated industries.

10.1 EXTERNAL ECONOMIES

Externalities are a primary subject of neoclassical economics and the new institutional economics. Externalities can be divided by their positive and negative impacts

[1]http://en.wikipedia.org/wiki/Business_cluster.

The Data Industry: The Business and Economics of Information and Big Data, First Edition. Chunlei Tang.
© 2016 John Wiley & Sons, Inc. Published 2016 by John Wiley & Sons, Inc.

as "external economies" (also called external benefits) and "external diseconomies." External economies, from the point of view of the enterprise achieving economies of scale and scope, are perceived to foster industry growth as well as its market efficiency.

The research milestones in this field due to Alfred Marshall's theory of value on external economies, Arthur Pigou's Pigovian taxes, and Ronald Coase's insight on trade of externalities captured by the Coase theorem. Marshall thought that external economies of scale cause the formation of industry clusters. Pigou supplemented Marshall's concept with "external diseconomies," and claimed that the government must levy taxes on negative externalities. The argument used by Coase to criticize Pigou was that voluntary negotiations could address external problems instead of Pigovian taxes. Theoretically, therefore, from the perspective of external economies, it is possible to understand the larger functions of data industry clusters, as well as their capacity to interconnect with external developments.

10.1.1   External Economies of Scale

External economies of scale are the basis for the relationship between production scale and average cost. When an enterprise experiences diseconomies of scale (because of increased market uncertainty impinging on its productive activities), the enterprise will want to spread its internal diseconomies outward; in other words, the enterprise will expand production to reduce average cost and increase returns. Let us say that the enterprise happens to be within a general industry cluster, and can leverage its costs by the geographic proximity of enterprises in the cluster; then the enterprise can expand production while not enlarging its own scale but establishing cooperative alliances and networks for sharing resources and costs. Consequently the scale of the industry cluster will increase, as can be measured by the increased number of same products.

Since there are no preexisting data industry clusters, the first problem we faced is whether the external economy of scale is the primary factor behind the formation of data industry clusters. The answer is negative, since data products that are customized to meet demand do not undergo large-scale production. However, we should never underestimate the corresponding effects, once a data industry cluster comes to fruition. The effects could be cross-field or cross-industry for a specific data product. Take a data product using transportation data as an example. Transportation data can significantly benefit medical, tourism, and logistics industries, such as planning ambulance real-time routes, mitigating overcrowded tourist attractions, and accelerating goods delivery. Likewise, a health early warning data product could boost the tourism industries.

10.1.2   External Economies of Scope

External economies of scope have been defined as "the economies of joint production and joint distribution" by renowned business historian Alfred Chandler [73]. Briefly, the connotation here is diversity – producing more than one product through specialization. That is to say, within a general industry cluster, intra-cluster enterprises can divide a production system into multiple parts through work specialization. They can cooperatively participate in the production of diversified products on a value chain, and build a cooperative network of frequent transactions, so as to cut innovation costs and reduce risks.

External economies of scope can indeed motivate the data industry to form an agglomeration. From the several enterprise agglomerations discussed in the previous chapter, external economies of scope had an inevitable role in the formation of the industry clusters. One reason is that intra-cluster enterprises' functional and hierarchical division of labor can target some common goals as a specific domain or sector data resource, and then engage in producing diversified data products. Another reason is future performances. There are no geographic limitations to data industry clusters; modes of enterprise agglomeration can even extend beyond the borders of a country.

10.2   INTERNAL ECONOMIES

Based on Marshall's theory, internal economies are standard measures of how efficient a single enterprise is in reducing costs or increasing revenue by its operations management, when it is enlarging its scale of operations. Here, we introduce two common standards, as shown in Figure 10.1, the horizontal standard indicates the number of duplicate products; and the vertical one, the number of production processes.

In an industry cluster, internal economies are mainly created by the internalization of intra-cluster enterprises' external economies. There are two general ways to do this. One is for a single enterprise to obtain added benefits after it has enlarged on its own scale. Second is for the single enterprise to acquire intra-cluster added benefits from all the value-adding activities along the value chain.

Given the fact that enterprises of data industry clusters are generally small in scale, and even of micro scale, the internal economies of data industry clusters may be obtained only by the value chain. Thus "coopetition" and synergy are the key means to obtaining added value.

10.2.1   Coopetition

The neologism "coopetition" describes both competition and cooperation, and has been re-coined several times. Adam Brandenburger, a Harvard Business School professor, and Barry Nalebuff, who teaches at the Yale School of Management, used this term in their 1996 book *Co-opetition*. Brandenburger and Nalebuff suggested that the relationships between the enterprises are not pure competition or cooperation, but actually a benefit game to pursue the dynamic balance between the contribution of capabilities and financial returns. This viewpoint has been accepted by a lot of people, and even Michael Porter added (1998) that industry clusters promote coopetition, and the coexistence of competition and cooperation "not only provides incentives but also avoids excessive competition" [74]. At current time, the intra-cluster coopetition relationship has extended to all of the interconnected suppliers, providers, as well as associated institutions that involves with cluster development.

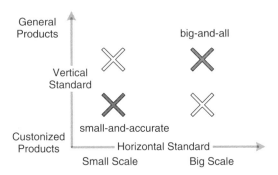

Figure 10.1   Measurement scale of an enterprise's efficiency

The general intra-cluster coopetition relationship has three characteristics. First is cooperation as prerequisite of coopetition. Intra-cluster coopetition under the premise of cooperation is complementary, such as launching a two-party or multi-party cooperative plan to complement each other's advantages, and promoting two-party or multi-party business performance by sharing skills or resources. Second is to look for win-win solutions in competitions. Despite the notion of competition as a win-lose situation in resolving conflict, the competitive sides can attempt to make some compromises in order to build on their intra-cluster coopetition relationships so that both sides gain and lose something. Of course, a benefit game always exists. Third is the market, which is the key to reconciling the contradiction between competition and cooperation, in that cooperation may enable a joint marketing venture despite the competition for market allocation.

Such an intra-cluster coopetition relationship, in the future data industry clusters, would need to include the following features: (1) external economic entities with similar work specialization; (2) data industry players and traditional industry players; and (3) unforeseen competition or cooperation, as various coopetitions intensify in times of increasing market uncertainty and risk.

Take the three types of enterprise agglomeration examples: driven agglomeration, industrial symbiosis, and wheel-axle type agglomeration. Human labor and scientific/technological forces will inevitably induce competition among multilevel data product providers and institutions creating data technology and industrialization of specialized work; however, competition for financial resources can be transformed into coopetition relationships for data industrial investment funds, multilevel data product providers, data resource suppliers, data industrial bases, intermediaries, and even local governments promoting regional economic growth, by the competitors complementing each other's advances. The entity symbioses, virtual derivatives, vertical leaderships, and growth poles will attract and retain the enterprises of other industries in the cluster and create a new coopetition relationship between data industry players and traditional industry players. Specifically, when data industry players assist traditional industry players in their upgrades and transformations, there is win-win cooperation; when instead data industry players attempt to seize the market, competition will result. This situation is shown in Figure 10.2, where the ✕ symbol

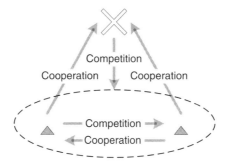

Figure 10.2   Coopetition between data industry and traditional industry players

represents traditional industry players, and the triangles denote data industry players.

10.2.2 Synergy

Synergy[2] means "working together"; synergy is a whole creation that is more than the simple sum of its parts, in that $1 + 1 > 2$. An industry cluster is envisaged as a system in which intra-cluster enterprises get more than just the profits through collaboration and more than they would in the case of individual operations. This is so-called synergy effect.

Compared with other industries, the data industry relies heavily on resources, technologies, and the capital. There are explicit specializations and also collective operations. As a result synergy effects of the future data industry clusters will not only have the general characteristics of other industries but be endowed with its unique features described below.

1. *Synergy Resources.*   Since data resources are the unique and core resource of the data industry, from the perspective of data possession, no enterprises can own a specific data resource or control it completely and permanently. Enterprises and other participants have to complement each other's data resources in a data industry cluster; and thereby enlarge or increase cluster overall revenues.

2. *Synergy of Systems.*   There is a common question on general industry clusters: how to build and sustain the culture, trust, and knowledge sharing. In the data industry cluster, the "code of conduct" will afford easy compliance, and meet the needs of all participants. For example, while customizing a data product, enterprises, institutions, and other organizations may be keen to reduce the frequency of meetings, and use much faster communications to share knowledge and innovative results, and even cut costs of the entire cluster by saving on several transaction costs.

3. *Synergy versus Agglomeration Economies.*   Often an industry cluster can help intra-enterprises form cohesive bonds by "risk sharing" in order to deal with

[2]http://en.wikipedia.org/wiki/Synergy.

fierce external forces. This also applies to the future data industry clusters. For example, in order to meet the interest demand of avoiding duplication of R&D activities and waste of data resources, intra-cluster data enterprises may need to change R&D from technological self-support to alliance cooperation in learning that another participant having more advanced data technologies is accelerating the R&D process, and this may be a university or a scientific institution. Sometimes influential data scientists, venture investors, intermediaries, industrial associations, as well as inspection and certification agencies can play positive roles and assist the data industry cluster.

4. *Resolve Conflicts in Synergy.* A good conflict solution requires effective implementation. In many cases this can help prevent cutthroat competition and facilitate reasonable competition. It is critical to promote fairness for all participants in terms of cost, price, technology, franchise, quality, talent, and management, and finally, to motivate latecomers to data innovation. Such coopetition can induce the competitive side to launch even more data innovation and to contribute to future interactive coupling and healthily developments within data industry clusters.

10.3 TRANSACTION COST

"Friction-free movement" is highly desirable in the physics community. Mainstream economists have likewise been attempting to reduce friction in transaction costs. And the industry cluster has proved to contribute to such reduction in costs.

The famous economists who have addressed the issues of industry clusters from the perspective of saving transaction costs, are the two Nobel Prize laureates in economics Ronald Coase (1991) and Oliver Williamson (2009). They argued that the industry cluster, as a special organization between a pure market and a hierarchy, can form a production cooperative network to reduce transaction costs and can protect cooperation by offering intra-cluster enterprises the opportunity to interact and enhance their creativity and flexibility.

Below we discuss the three main theoretical insights of Coase and Williamson and analyze the transaction-cost effects of the data industry cluster: the division of costs, opportunity costs, and monitoring costs.

10.3.1 The Division of Cost

Although Coase and Williamson take somewhat different routes, they both contribute an important insight to transaction costs that are implicit in labor specialization (or division of labor) and complementary as well.[3]

The notion of intra-cluster data enterprises being small but excellent and fast but accurate may be due to five advantages of labor specialization within a data industry cluster, which a single data enterprise would not experience.

---

[3] http://hes2011.nd.edu/assets/42967/b_ylund.pdf.

1. *Independence from Administration Costs.*   Within an enterprise, the division of costs included in administration costs as well as the organizational size depends on internal expansion or contraction. They have been shown to constitute a comparatively large proportion of administration costs. In comparison, within a cluster, the division of costs is independent from administration costs; moreover transaction frequency, or what is considered "high-frequency" within the cluster, can further reduce the costs of specialization.

2. *Team Selection.*   Within an enterprise, labor specialization reflected in one's occupational concentration in a specific area of expertise is a personnel selection. In comparison, within a cluster, intra-cluster institutional arrangements are a team selection. Therefore parties to a transaction can clearly express their thoughts, negotiate and detail their rights, responsibilities, and obligations in strictly executed contracts.

3. *Complementary.*   Intra-cluster enterprises are judiciously distributed along a value chain, to facilitate their complementing each other's advantages and to further cut costs. In contrast, such complementary rarely occurs among non-cluster enterprises.

4. *Sharing Resources (Specially Data Resources).*   Intra-cluster resources have exclusive ownership, and often the benefits derived from such exclusiveness promote sharing among intra-cluster enterprises to control the price mechanism. Specially, the resources sharing includes data resources.

5. *Asset Share.*   Despite there not being much infrastructures in the future data industry clusters, sharing underused assets can easily relate to intra-cluster enterprises breaking down the limitations of asset specificity, in order to further reduce the cost of unusually idle expensive facilities and equipments. The more enterprises use intra-cluster facilities and equipment, the lower assets' idle time and costs (relatively speaking).

10.3.2   Opportunity Cost

Coase discussed the difference between cluster effects for cost and enterprise regulatory structures, which is the size boundary. An enterprise has an organizational size; however, the industry cluster is quite different as it has no "boundary." Managing the allocation of resources for an enterprise is an administration cost; the specialization costs depend on the organizational size, whether it is big, medium, small, or micro-size. To get high administration costs in line, an enterprise could enlarge its organizational size. When the administration costs equal the transaction costs on the market exchange, the alternative market commitment will end. In contrast, the industry cluster is not limited to administration costs; its larger size tends to have greater external effects. It was this concern that led to Coase's insights on the importance of thinking in terms of opportunity costs[4] where organizational size is involved in the subjective choice of enterprises.

The opportunity cost, also known as the alternative cost, consists of two opposite parts: implicit cost and explicit cost. Implicit cost is the opportunity cost that is equal

---

[4]http://www.gredeg.cnrs.fr/colloques/gide/Papers/Elodie%20Bertrand.pdf.

to what a firm must give up in order to use a factor of production that it already owns and does not pay rent for[5] – the value of the best alternative forgone; while explicit cost is the monetary cost paid to resource owners.

Compared with intra-cluster and non-cluster enterprises, intra-cluster enterprises have lower opportunity costs. One reason for this is that intra-cluster information dissemination is much faster and more thorough because intra-cluster enterprises have more opportunity to communicate with each other as to their specializations and also more coopetition, so the information asymmetry that could result from the increased opportunity cost is mitigated. The second reason is that, thanks to bounded rationality, intra-cluster enterprises are prone to sign simple contracts or agreements, which is not the case for non-cluster enterprises. Generally, non-cluster business activities are configured and coordinated by the markets or several market-leading enterprises; there may be trust but without dependence in that contracting parties do not trust each other. Therefore non-cluster enterprises might only provide for and understand the future under bounded rational conditions, and are prone to draft complicated contracts with detailed clauses that may be unfavorable for the other party.

10.3.3 Monitoring Cost

Williamson divided costs into arising prior to and after the moment of transaction. Ex ante costs are used to identify the interests, obligations, and the rights and responsibilities of the parties involved in a transaction; and a price to pay for such a process of identifying is related to the clarity of the ownership structure for both sides. Ex post costs are further divided into several phases as administrating, monitoring, and enforcing, which all involve enhancing implementation. Enhancing implementation capabilities often means the processes that are in place for monitoring the status, risks, costs and benefits of implementation activities; thus in normal circumstances the lower monitoring costs pay the better per completion.

In a general way the monitoring cost are most prominent in the large business, and can rise along with increases in organizational size. This is because a large enterprise employs more people, and the interaction among employees will increase accordingly, require additional coordination work, and thereof be intensified. For example, if a company has four employees, it may assign two additional employees to coordinate the relationships among the four; and the formula for coordinating such an interaction is $n(n-1)/2$.

If a similar situation occurs in an intra-cluster enterprise, the solution is relatively cheap. The reasons are that (1) due to coopetition maintained by the intra-cluster price mechanism, enterprises have no need to internalize production activities; (2) diversification can prevent excessive competition, since intra-cluster enterprises prefer to use different resources or sets of resources engaging in their production activities; and (3) intra-cluster enterprises often increase transaction frequency to extend their cooperation.

In brief, for the future data industry clusters, real-time monitoring of other enterprises will be unnecessary during the supply of diversified data products because

---

[5]http://en.wikipedia.org/wiki/Implicit_cost.

overall production activities are controlled by the enterprises of the data industry chain. The diversified data products involve just two functions: (1) innovations that support the breadth of application and (2) collaborations that underpin the depth of application.

10.4   COMPETITIVE ADVANTAGES

Michael Porter, in explaining his diamond model, dispelled the traditional David Ricardo theory of comparative advantage, in that it fails to account for the source of industrial competitiveness. Soon after that, in November 1998, Porter published his paper "Clusters and the New Economics of Competition" in the *Harvard Business Review* [74] in which he argued that "clusters affect competition in the three broad ways: first, by increasing the productivity of companies based in the area; second, by driving the direction and pace of innovation, which underpins future productivity growth; and third, by stimulating the formation of new businesses, which expands and strengthens the cluster itself."

In this regard, from the perspective of competitive advantage of the emerging data industry clusters, there may be ways that the clustered environment can facilitate the emergence of new intra-cluster enterprises, then form and extend cluster effects, so as to "influence on productivity and especially on productivity growth" [75].

10.4.1   Innovation Performance

In recent years, under the continual influence of evolutionary systems and organizational learning and innovation systems, there is a growing consensus that innovation is an evolutionary process involving individuals and groups. A special study[6] implemented by the United Nations Industrial Development Organization (UNIDO) found that intra-cluster structure is especially significant for innovation growth in developing countries. This is because only a small part of what constitutes knowledge management in innovation projects is derived from sources within an enterprise; the majority of critical innovations cannot be done by a single enterprise alone. Cluster effects on innovation performance are reflected in supply and demand. For the supply side, it can directly facilitate work specialization, and indirectly accumulate knowledge for a specialized field. For the demand side, it can help establish market advantages in a sector, and naturally leads to scale effects.

AnnaLee Saxenian's findings [76] and Rui Baptista [77] provide strong evidence that clusters are the drivers of enterprise innovation and they motivate innovation by their successful performances. Saxenian is a professor at UC Berkeley's School of Information. She is known widely in her original research of the Silicon Valley technology cluster, which is very common to solicit help from a competitor to tackle a technical challenge, and the counterpart enterprise is generally willing to help. Such an exchange has gained intra-cluster enterprises a unique advantage in innovation.

---

[6]https://www.unido.org/fileadmin/user_media/Services/PSD/Clusters_and_Networks/publications/ceglie_dini.pdf.

Baptista, who comes from Carnegie Mellon University, has argued that cluster effects shorten the feedback loops in innovation, so that enterprises do not need bear all the financial risk.

Similar relationships can be expected in the future data industry clusters. Likely innovation will depend on comprehensive foregoing analyses, and data resource developments, in multiple stages. (1) In the first stage. A certain traditional enterprise would be willing to use its owned or controlled data resources, and seek to establish cooperation with some known universities. Any research Institutes dedicated to data technology and its industrialization at these universities would be the main focus. (2) In the second stage. Under the guidance of venture capitals, some enterprises or individuals are likely as potential bidders to participate in bid submission of a data prototype produced by above-mentioned institutes, and then will transfer the data prototype into a real data product. Some intermediaries might intervene at this stage in the form of dissemination organizations, industrial alliances, and training agencies. (3) The third stage. The traditional enterprise will try to use the data product, and this may lead to product improvements or application innovations. (4) In the fourth stage. The data product would be carried over to other multi-domain and cross-sector projects to form external economies, thus drive a new round of transboundary cooperation. Taken as a whole, if timely adequate funding is obtained from venture capitalists, all these stages could produce intersections and even overlaps, and eventually contribute to more innovation from positive feedback.

10.4.2   The Impact of Expansion

John Henderson [78] from Brown University was an early researcher on the contribution a dynamic cluster can make to regional economic growth, and the key measure is its scale effect (and another is the detailed division of labor). Therefore the impact of expansion is a unique competitive advantage that an industry cluster develops external economies over the short term to drive the swift growth of local, regional, or national economies.

The impact of cluster expansion can be perceived from two dimensions: the growth of a single enterprise within a cluster and the cluster holistic expansion. In the case of the growth of single enterprise, it is the conscious strategy choices that the enterprise makes in seeking its own advantages for efficient resource allocation, which include multiple methods of enterprise growth. Compared with other methods of enterprise growth such as general growth, M&A growth, and cross-organization growth, clustering growth is that obtained or mobilized as an enterprise's resources become acquired more externally. In the case of the cluster expansion, this can generally be classified into (1) lateral expansion, which allows investors to easily find market opportunities without risk concerns, and can increase the number of new enterprises within the cluster; and (2) vertical expansion, which allows an enterprise to expand toward both the upstream and downstream enterprises, and attract substantial other enterprises to provide services.

For the future data industry clusters, there can be anticipated three economic impacts due to expansion: (1) cluster effects of scale resulting from the accumulation of innovations, (2) cluster effects of diffusion resulting from advancements in data techniques, and (3) cluster effects of scope resulting from capital aggregation.

10.5 NEGATIVE EFFECTS

An industry cluster has, no doubt, some negative effects, which are usually offset by the positive impacts mentioned above. Representative cases of a negative cluster effect in the strategic convergence of the value chain are the declines of the Detroit automobile and the Qiaotou button industries, both of which lost their competitive advantages. As Edward Glaeser [79] of Harvard has said, heavy dependence on an individual industry can severely weaken its diversified competitive advantages, and cause a cluster facing external threats or internal rigidities to lose its competitiveness. From the viewpoint of externalities, negative effects usually are detectable in the cluster's life cycle during its dynamic evolution, and may be pronounced at different stages. When a cluster considers accelerating its economic growth, it is may be wise to recognize the possible negative effects.

With the data industry cluster still at the early stages of its life cycle, the negative effects could show up as an innovation risk, a data asset specificity, or a crowding effect.

10.5.1 Innovation Risk

In his best known 1962 book *Diffusion of Innovations*, Everett Rogers [80] introduced five channels of innovation diffusion: by innovators, early adopters, early majority, later majority, and laggards. He argued that market acceptance of innovation is not a continuous process. In his opinion, early adopters represent the early-stage market, and later majority represents the mainstream market, while the major difference between the two creates a so-called gap, namely risk. In this regard innovation risk has three characteristics: (1) external uncertainties, (2) difficulty and complexity of innovation project itself, and (3) limitations of the internal resources and innovation capabilities.

For the data industry, innovation risk is mainly caused by not understanding the data technologies. On one hand, it is difficult for innovators to clearly describe an application scenario to early adopters before the actual implementation; for instance, description of big data as "sampling" tends to have larger implications and even the "total sample" results are approximate rather than exact, such that correlation overweighs causality. All this can mislead early adopters to conclude that big data results cannot be verified. Yet, despite not having participated in any innovation activities, early adopters tend to think themselves to be enough informed to educate the early majority, thus further exacerbate the gap. For example, a person could be someone who never actually implemented data mining but knows something only about data analysis, and instructs other persons how to change thinking in response to the big data reform.

In brief, in the early stages, innovation risk can hinder the growth and development of data industry clusters, and may even increase the risk of premature mortality.

10.5.2 Data Asset Specificity

Ideally, in the early stages of clustering, intra-cluster enterprises agglomerate by longitudinal specialization along the industrial chain. Such a linkage-type dependence of the upstream and downstream may nevertheless include enterprises that are unwilling to share their assets. Such a negative effect as asset specificity among intra-cluster enterprise can weaken resilience to external challenges. Thus the failure of a link on the industrial chain can have a domino effect and jeopardize the entire industry cluster.

Asset specificity is likely to be particularly harmful in the data industry clusters. This is because data enterprises have a special asset that other industries do not have: the data asset. The data asset specificity can lead enterprises to competing for data resources. Putting data assets under the price mechanism may lower and even eliminate such a negative effect.

10.5.3 Crowding Effect

The crowding effect usually appears at the mature stage of a cluster. As a cluster matures, increasing knowledge spillovers and diffusion can result in the cluster easily achieving external economies and lead to innovation inertia. The innovation stagnation in the entire cluster could then result in product duplication and cut-throat competition within the cluster.

Likely, for the data industry, the crowding effect will emerge in the early stages. The signs to look for are big data hype, government encouragement, and capital directional guidelines, as these developments can cause individual enterprises to swarm and overcrowd. If additionally a large number of enterprises wholly focus on several specific data resources, or traditional enterprises, data technologies, there will definitely result disorder competition among intra-cluster players, with product duplication and market crowding out. All this damage to the interests of all players can further hinder the normal development of the data industry cluster.

11

A MODE OF INDUSTRIAL DEVELOPMENT FOR THE DATA INDUSTRY

In Alfred Marshall's industrial organization theory, and subsequently that of Barry Weingast, the mode of industrial development is essentially a bureaucratic system that provides for government interventions (e.g., in various levels of planning) from an economic policy perspective. The mode of industrial development specifies the industrial structure, industrial organization, industrial distribution, industrial strategy, and industrial policy. However, such a top-down system, if inappropriately designed, can interfere with the basic operations and vicissitudes of a developing industry, and ruin the industry and even cause it to collapse.

11.1 GENERAL ANALYSIS OF THE DEVELOPMENT MODE

The development mode is presently a research hotspot in both development economics and industrial economics. It is used to indicate future direction for industry, country, economy, or society. The mode takes into account the history, culture, and existing resources, and thus is designed (or selected) according to the unique characteristics of a particular industry scenario.

Needless to say, there are no good or bad development modes; there is only the most appropriate mode for the specific situation in a certain site. For example, when considering creating a site, we should take into account whether it will fit the current situation, provide a unique advantage, and contribute to sustainable growth.

The Data Industry: The Business and Economics of Information and Big Data, First Edition. Chunlei Tang.
© 2016 John Wiley & Sons, Inc. Published 2016 by John Wiley & Sons, Inc.

11.1.1 Influence Factors

There are many factors that go into planning a development mode. On the one hand, there are commitments to include resource endowments, strategies, and the innovation area; on the other hand, there are the external commitments to global standardizations, and to then local policy regulations in finding the quintessential locale.

1. *External Environment.* The exact nature of the immediate surroundings is used frequently in determining the development mode choices. The natural resources, other industry, the country, the institutional system, cultural values, and the diversity, centrality, and stability of the people are all factors that can influence the choice of building site.

2. *Natural Endowments.* Natural endowments may provide the motivation for developing a site. But there are the economic considerations of natural endowments maximization, and control the flux and flow direction of complementary factors. For example, in building a mode of industrial development, it is important not only to consider the casual relationship, hierarchical relationship, and functional relationship, but also to study coupling effect and input proportions of variable factors, so as to be ready to translate replacement competition into complementary cooperation.

3. *Diffusion of Innovations.* Diffusion of innovation, as argued by Everett Rogers, is "the process by which an innovation is communicated through certain channels over time among the participants in a social system." To some extent, forethought as to the rate of diffusion can determine the design and selection of a development mode. This would include the technological system, technological readiness level, core technologies' control, and extension possibilities of the technologies.

4. *Policy Guidance.* Various regulations due to government policies can obviously constrain the development mode. But government can also reallocate capital and resources, provide guidance as to the direction of development, and remove barriers from the external environment.

11.1.2 Dominant Styles

Theoretically, there are two dominant styles of the development mode – government and market. These are relatively simple to map out.

A government is dominant in third-world countries intent on developing an advanced market economy. In the eighteenth century, Friedrich List, a forefather of the German historical school of economics, advocated government support of the economics of nations in a political union with a unified domestic market, which was then a contrary position for Germany, which was inclined to follow the actual practice (like the British and French cases) or rather the abstract doctrines of Joseph Smith. Likewise, after the Second World War, Japan and other countries of Eastern Asia adopted this dominant role of government.

In today's world, there is no pure dominant government style. This is because all the development modes that are initiated and developed by governments can be traced back to government intervention, even if intended to be an "invisible hand" based on Adam Smith's idea of trade and market. For instance, the natural growth model of the US information industry is still controlled by the government, despite being all the while based on consumer demand and free market adjustments.

11.2 A BASIC DEVELOPMENT MODE FOR THE DATA INDUSTRY

There are two ways to design a development mode for an emerging industry. The difficult way is to address the characteristics of the industry and take into account all the factors that can predict its success. The other is to reference or draw on the existing experiences of similar industries, which is relatively easy.

As discussed previously, the data industry is the reversal, derivation, and upgrading of the information industry. The information industry has already built a basic development mode for the data industry, and the reference countries to consult are the United States, Japan, South Korea, and India. Among them, the United States has good natural growth model determined by the market; Japan and South Korea have selected the typical East Asian model dominated by government, and India's software outsourcing model has been developed by its government.

11.2.1 Industrial Structure: A Comprehensive Advancement Plan

Breakthroughs in the industrial development develop comprehensively or in discrete key areas. The information industry in the United States has developed comprehensively with the hardware manufacturing and software service sectors having advanced side by side. Today, American's hardware and software development activities are both leading the world. Japan, South Korea, and India have taken breakthroughs approach in some key areas. Japan has focused on the "bigger and faster" development, such as computers for large-scale and high-speed applications and semiconductor chips for miniaturization and high-capacity. South Korea has focused on the development of semiconductor memory, application-specific integrated circuit (ASIC), liquid-crystal display (LCD), and mobile devices. India has engaged its breakthroughs only in software services.

Before beginning the mode design, it is important to study the locale's industry sector regarding the operations and relationships within the existing industries. This is because (1) the present socioeconomic foundations create the conditions for facilitating the development of an emerging industry and (2) the structure with its inherent relations of existing industries can help identify future development priorities and sequences for a new industry.

Since data resources spread the world over, the plan adopted should be comprehensive and designed to advance the data industry structure. More specifically, there are three interdependent levels of such a structure: first, at the macroscopic level, the mode will likely be subjected to government regulations, among other things, in relation to formulating a national strategy for developing the data industry; second, at the mesoscopic level, the mode will likely attract venture capitalists' and other enterprises' investments, leading to clustering based on contributions and adjustments by traditional businesses, upgraded consumption configurations, and improve cultural environment; third, at the microscopic level, the mode will likely encourage data innovations at R&D institutions and open source development, and thus sharpen market competition for the survival of the fittest.

11.2.2   Industrial Organization: Dominated by the SMEs

The market structure, in which a number of firms are producing practically identical products,[1] is an important criterion in industrial organization. The sizes of producers in the information industry are big-business enterprises and the SMBs.

In the US information industry there is implemented a combined strategy of competition and monopoly. In other words, the US government has offered, on the one hand, positive support to the SMBs through venture capital investments and the development of secondary markets to encourage perfect competition. On the other hand, government has encouraged large enterprises to participate in international market competitions and thus supported frequent restructuring through mergers and acquisitions. Japan and South Korea have pursued a "vertical" domination by large business groups – specifically, by Korean family-owned conglomerates called *chaebols* and by Japanese conglomerates known as *keiretsus*. In both Korea and Japan these associated the SMEs take the form of the point-axle type agglomerations, in which the SMBs that supply the primary technology spill over products to big-business enterprises at the top; otherwise, the SNEs do not have a good chance to survive. The sizes of enterprises in India are also large, the main ones being the Tata Consultancy Services (TCS), Infosy, and Wipro. Wipro, for example, has acquired seven US-Australian businesses at a cumulative cost of US$200 million only in two years from 2003.

Before beginning the mode design, fully two things need to be considered: the means for the enterprises to participate in the market, and the connections between the enterprises of the data chain and external policies. These are the very two things that can directly influence the direction and indirectly determine the positioning of an emerging industry.

A big-business enterprise of industrial organization is suitable for the information industry, due to its hardware manufacturing division, which has the basic principle of "bigger and stronger." In comparison, there is no such division in the data industry, and thus no need to copy this principle in the emerging data industry. Industrial organization of the data industry is better tuned to the SMEs to leverage a rapid response mechanism.

11.2.3   Industrial Distribution: Endogenous Growth

Industrial distribution is somewhat like industrial transfer. When we study a region's or a country's industrial distribution, we are actually researching the inevitable

---

[1] http://en.wikipedia.org/wiki/Market_structure.

post-transfer changes in the industrial structure based on two growth models: endogenous dynamics and exogenous forces. In general, endogenous dynamics are the result of agglomerations of intra-industry enterprises driven by complementary industries. Exogenous forces result from industrial transfers or direct investments.

In the information industry, the United States attaches great importance to industrial clusters, such as California's Silicon Valley and Massachusetts' Kendall Square. Both areas have shaped the endogenous dynamics that play to the concentration effects of industry clusters. Due to limited domestic resources and markets, Japan, South Korea, and India have all adopted a globalization strategy that depends on industry transfer to absorb new knowledge and technologies, and thus accelerate their industry sectors' growth.

From the beginning, planning of the development mode has to address issues of data resource allocation, configuration, and utilization. If we decide to encourage industry transfers, the mode for the data industry will to some extent be spearheaded by the leading enterprises. The industry transfers must be selectively undertaken, since some such enterprises may transfer limited knowledge or technologies and aim merely to expand their market shares.

Based on the readiness of data technology, the East-West gap has narrowed enough to disregard the notion of industrial transfer. Many top data scientists are Asians, and among the best are especially the Chinese, Indians, and Japanese. Therefore, industrial distribution of the data industry that should be adopted is an endogenous dynamics growth.

11.2.4   Industrial Strategy: Self-Dependent Innovation

Product innovation often is not radical but incremental. Yet, the innovation process can vary according to the particular innovation type. From this perspective, there are four possible ways to strategize industrial R&D: as unique, integrated, amalgamated, absorbed, or collaborative innovation. The first two are essentially self-dependent innovations.

In the information industry, the United States emphasizes self-dependent proprietary innovation for both fundamental research and major breakthroughs. These are time-consuming undertakings, and for developing innovative projects, large laboratories have been constructed and big investments made on high-risk research. The steps of basic research as applied to the technology market's development have enabled the US technology enterprises to naturally evolve. Japan, South Korea, and India have all instead focused on projects that yield quick returns with little investment, and have amalgamated and absorbed their innovations.

It should be remembered that data science is an applications-oriented technology. The data serves to improve and refine present-day socioeconomic circumstances, so the research-development path is not long. Likely, early interventions of venture capitalists could nevertheless help bridge the gap between basic research and market development in order to enable the data industries there to launch unique and integrated self-dependent innovations.

11.2.5 Industrial Policy: Market Driven

Industrial policy is all about setting up and enforcing codes, standards, and recommended practices for conducting legitimate markets and economic activities. There are two types of government regulations: state driven and market driven.

In the information industry, the United States combines national macroscopic demand management with laissez-faire. More specifically, in the early stage, the government directly intervenes by means of government procurement and investment; in the intermediate stage, the government gradually turns to indirect regulation of policy formulation; in the late stages, the government mainly targets in antitrust investigations. Government economic policies in the East Asian model adopted by Japan and Korea is more comprehensive with in-depth macro- to mesoscopic government intervention, from strategic planning and legislation to institutional coordination.

Before design planning can begin, there has to be identified the extent of government economic intervention, which generally does remain consistent. This is because the policy that has relative stability is the preferred policy for an industrial development, as opposed to a continually evolving policy.

The data industry has the advantage of being an upgrade of the information industry. It can fall in with the pathway of government gradually decentralizing power to the market mechanisms in allocating societal resources, and thus be market driven though supplemented by government regulations.

11.3 AN OPTIMIZED DEVELOPMENT MODE FOR THE DATA INDUSTRY

Attending only to the industry differences between the information and data industries in building a model for the data industry is not a realistic approach. The innovative approach is to learn from the defects of the development mode of the information industry, and to rectify them according to the mode if they are contrary to the characteristics of the data industry.

11.3.1 New Industrial Structure: Built on Upgrading of Traditional Industries

From the perspective of the industry sector, globalization is remaking industry economies and this is an inevitable challenge for the data industry. The sooner the data industry prepares to face the challenge, the easier it will be to overcome. For the data industry the challenge is to optimize the industry structure in terms of the regional economy as well as to stagger development among all regional economies. One option may be to upgrade or transform the traditional industries, and another may be to make an industrial gradient transfer. Next would be to determine how such choices could serve the data industry itself.

The theory of industrial gradient transfer, proposed in 1962 by the 2012 economics Nobel laureate Vernon Smith (2002), advocates that the industries of more advanced regions that pioneer the development of leading-edge products should cross over to less advanced regions and thus drive the development of the overall economy. There are nevertheless a couple of obvious obstacles in actual practice. One is that the industries that will transfer often are a detriment to the more advanced regions because they are high polluters, with high carbon emissions, high energy-consumption, low-tech, and low value added. The other is that the gradient transfer often involves scrambling the headquarters. In other words, to retain a competitive edge, the output region will nominally retain a headquarters with many preferential policies, though in the initial years, the input region may fail to get the corresponding financial returns from headquarters after paying additional externalities costs, for example, in cases of pollution that devalues environmental and natural resources.

At an operational level, industrial upgrading, as defined by Gary Gereffi, will first at the microscopic level enable economic entities to move from low-value to relatively high-value activities and transform the industrial landscape such that then at the mesoscopic level the change to the production and consumption structures will lead to environmental sustainability [81]. Thus, on the one hand, by the upgrading and transformation, business process management is optimized without overhauling the core businesses, and on the other hand, total restructuring is based on acquired new resources or capabilities. In the first case, the original industrial foundation has a relatively low withdrawal penalty; in the second case, withdraw from the original industry comes with relatively high economic cost and time, as well as entry costs into a new unknown territory.

In any case, a comprehensive plan is critical in developing the data industry. As mentioned above, the industrial layout could be integrated by transferring, updating, and transforming traditional industries, and then evaluating the impact and effectiveness of innovative collaboration. We further believe that industrial updating with data innovative interventions would be especially effective in China. In Chinese manufacturing, goods such as home appliances, clothing, and accessories, and also chemical engineering are indispensable to global trade. With innovative data advancements, all such traditional manufacturing industries could be upgraded to attain "win-win" productivity levels.

To be sure, a new industrial structure must deal with issues of native culture, especially foreign cultures. World cultures are diversified, and a global scale monoculture due to a homogenization of cultures is akin to worldwide cultural decay.[2] Each unique regional culture is influenced by all participants in the internal market, by the demographics, lifestyle shifts, economic cycles, and even the overall economic, regulatory, social, and political conditions. Unfortunately, one of the failings in the economic globalization tide is that multicultural coexistence has been threatened, and mainly by industries failing to understand regional resource endowments, the real needs and hopes of native people, and existing local industrial structures. It may be possible to address some of these difficulties using data technologies in place of conventional tactics. The conventional approach has been to employ census statistical techniques, but local governments that allocate substantial human and material resources from routine administrative funds are unwilling to do this at such large-scale as data analysis.

[2]http://en.wikipedia.org/wiki/Cultural_diversity.

11.3.2   New Industrial Organization: Small Is Beautiful

Industrial organization has historically focused on perfect competition in an idealized market. Of course, the idealized market does not include government incentives such as tax credits, assistance, and subsidies, since fairness and flexibility cannot be assured in government programs. As a result economic analyses at the micro level must rely more on market incentives to determine enterprise scale and capacity, and ensure free competition, that can stimulate an emerging industrial development.

The phrase "small is beautiful," in contrast with "bigger is better," came from a 1973 collection of essays by British economist Ernst Schumacher. This phrase is often used to champion small, appropriate technologies that are believed to empower members of the general public more.[3] Schumacher's economic thoughts generalized in one sentence is "an entirely new system of thought is needed, a system based on attention to people, and not primarily attention to goods – production by the masses, rather than mass production" [82]. He foresaw such "production by the masses" to meet three conditions: small-scale use, low prices, and cater to the needs of people. Coincidentally, the data product does satisfy all three of these conditions, and the data even caters to the special needs of very few consumers, that is, the niche productions in which generally large enterprises are unwilling to engage. Only groups of the SMEs and micro-enterprises can meet the objectives of the data industry. This is because the SMEs and micro-enterprises can do a fast launch, have low development costs, and can easily get venture capital funding.

11.3.3   New Industrial Distribution: Constructing a Novel Type of Industrial Bases

Industrial "clusters," or geographical concentrations of enterprises and ancillary units engaged in the related sectors, have been around for a very long time, serving as a catalyst for worldwide economic growth; in particular, for the US information industry. Think: Silicon Valley, Research Triangle Park, Wall Street, and even Hollywood. Besides this, there are the numerous outgrowths of industry clusters with a special themes, such as industrial base, business park, science park, hi-tech park, university research park, technopolis, and biopark, all of which depend on an affiliation with a university or a scientific institute, with the sort of science and research in which the cluster's entities engage.

It may be that for developing the data industry; a new concept engages with or influences data industrial bases, which could be developed with the following features:

1. *Cluster Cohesion Is Obvious.*   Although an industrial base might not include an entire industrial chain of the data industry, it can support competition and cooperation among enterprises within the cluster, sharing data resources, developing data technologies, producing data products, and so forth.

---

[3]http://en.wikipedia.org/wiki/Small_Is_Beautiful.

2. *Driven by Both Capital Investment and Data Innovation.* Instead of being driven solely by science and research in a traditional way, data industrial bases can, under the direction of the specialized venture capitalists, make arrangements with universities and scientific institutions to participate in their innovations.

3. *Humanizing Management.* Humanizing management does not mean laissez-faire. The primary purpose of data industrial base management is not just to run realty management system in a secure, controlled, and repeatable free-space environment but to uphold an accommodating working environment.

4. *Very Diversified Ecological Community.* A data industrial base is substantially a central business district embedded in the core functions that lead to data innovations, and a livable ecological community provided with commercial facilities and all-around cultural amenities that enhance creative work and routine activities.

5. *Multilevel Talent at New Heights.* A data industrial base should be capable of sourcing, attracting, selecting, training, developing, retaining, promoting, and moving employees through the organization. This involves different professional disciplines at all levels; thus, in pushing up morale, can rouse the staff's creativity via their enthusiasm, and with generous remuneration, including position matching and academic upgrading, can make work a cozy life.

11.3.4 New Industrial Strategy: Industry/University Cooperation

For an emerging industry, from the perspective of industrial strategy, a support system has to be established for innovations early on. Generally, constructing a support system is complicated, but for the data industry, there is fortunately a shortcut to success, that is, establishing industry/university cooperation.

Partnering between university and industry can be traced back to UNESCO's university-industry-science partnership (UNISPAR) program, which was launched in 1993 to encourage universities to become involved in industrialization. In the past several decades, the industry/university partnership has been widely recognized as a major contributor to the US industry developments. However, problems for both participants in this cooperative venture do exist, mostly due to the fact that traditional funds only support discrete research projects, piecemeal business ideas, and short-term or insubstantial investments. In China, for instance, over the years, economic growth has rarely depended on independent innovations. The situation is particularly serious in the following areas. First, due to lack of mutual understanding, the rights and obligations of both universities and enterprises are unclear. Second, universities often cannot advance researching methods, with the efficiency and speed that businesses require, causing enterprises to be edgy on their side because they perceive universities to be leisurely on the other side. Third, research outputs from universities do not meet the actual needs of enterprises; universities often are more interested in disseminating and publishing their research findings and filing for patents or copyright, all of which is far from commercialization. For the few companies that thirst for new technologies and new products and do get some real "guidance" from universities, the university investments often overlap and the rush into hot technologies generally leads to vicious competition.

To adequately develop the basic model of the data industry, the existing university–industry partnership needs to be improved. This may need to include, but are not limited to, [83] (1) intervention by policy makers or other facilitators, (2) intervention by the funding source that not only finances but captures its benefits, (3) intervention by scientific research publishers who could transfer findings to industry and become contributors to the growth of the economy, (4) intervention by intellectual property offices of university research that can impede or encourage commercial development, and (5) intervention by innovators of multiple identities in different economic entities. In fact data technologies can be applied to amicably getting the needs of all sides and even third parties in the open. It may even be possible to get the existing industry/university partnership to address demand-driven proprietary innovation within the data industry, and to globally expand the existing partnerships with the help of the data industry itself.

An example of a new approach to industry/university cooperation is being pioneered by InfinityData Investment Co., Ltd. The aim is to remake industry/university partnerships and guide universities to actively participate in data industrialization, using data innovations to meet the actual needs of all parties, so as to realize commercial potential. InfinityData maintains a specialized data industrial fund for this purpose headquartered in Shanghai, China. This is primarily to invest and establish a couple of institutions, based in data science and its industrialization combines broad knowledge across a spectrum of disciplines, which are affiliated to each of the world's top universities. Its secondary aim is to purchase multi-domain data resources that temporarily focus on healthcare, transportation, and finance. A third aim is to construct novel type data industrial bases in the first- and second-tier cities of China, namely Beijing, Shanghai, and Wuhan. A fourth aim is to serve as a resource for young people interested in entrepreneurship and data innovation, such as providing business ideas and related data technologies, relieving office space rent or reducing electricity tariffs, and being a consultant on routine operations. A final aim is to help with the upgrading and transformation of traditional industries that own or control data resources.

11.3.5   New Industrial Policy: Civil-Military Coordination

Covering almost all profit-making activities and the various data resources still under development, whether civilian or military, is an essential condition for operational autonomy and information superiority. The data industry has the military and civilian overlap capabilities to achieve tactical, operational, and strategic objectives, and challenges of the data industry are internationally capable of stimulating countries worldwide to explore the possibilities of reaching their competitive advantages.

To acquire an appropriate civil-military model for the data industry, there are to be included, no doubt, issues of civil-military coordination that guarantee effective dual-use coordination and the most efficient capability development for national defense forces. Developing the data industry may in fact prove to be the only harmonious way for countries to gain strategic advantages in the incessant rounds of international competition.

12

A GUIDE TO THE EMERGING DATA LAW

The data industry is principally making profits from the use of data. This industry consists in substantially a wide range of profit-making activities spontaneously carried out by cluster enterprises. The enterprises all gain from innovations within the data industry chain system, but so far there are no corresponding laws and regulations to compel them to implement risk prevention strategies and data resource evaluations, and to assume legal responsibility for infractions.

Even though there are many specific international and domestic regulations concerning abuse of data; a data legal system has not yet been formed. This chapter offers a development perspective for the data industry to discuss resources rights protections, competition institutional arrangements, industrial organization regulations, and financial supporting strategies, so as to facilitate corresponding legal system innovations in the jurisprudential circle.

12.1 DATA RESOURCE LAW

The social norms during a data-driven commercialization process are reasonable data resource sharing and care in preventing data resource abuse. Related laws and regulations can be traced to acts of information sharing and computer anti-abuse.

In regard to information sharing, the United States has adopted very different attitudes toward state-owned versus private information. State-owned information is completely open to the public, whereas the private information is well protected. The relevant federal laws include (1) the Freedom of Information Act (1966), the

The Data Industry: The Business and Economics of Information and Big Data, First Edition. Chunlei Tang.
© 2016 John Wiley & Sons, Inc. Published 2016 by John Wiley & Sons, Inc.

Privacy Act (1974); (2) the Government in the Sunshine Act (1976); and (3) the Copyright Act (1976). The Freedom of Information Act is often described as the law that allows public access to information from the federal government. The Privacy Act regulates the behaviors of federal agencies that govern the collection, maintenance, use, and dissemination of personally identifiable information about individuals maintained in federal records. This act provides open access to the individual to whom the record pertains and prohibits disclosure of this information to third parties. The aim of the Government in the Sunshine Act is to assure and facilitate the citizen's ability to effectively acquire and use government information. The Copyright Act explicitly stipulated in Section 105 that the federal government is not allowed to have copyrights, and there shall be no restrictions on reusing data for derivative works. The European Union has developed more comprehensive and systematic legal codes including (1) the Directive 96/9/EC of the European Parliament and of the Council of 1996 on the legal protection of databases, (2) Regulation (EC) No 1049/2001 of the European Parliament and of the Council of 2001 regarding public access to European Parliament, Council and Commission documents, and (3) the "Bucharest Statement" of 2002. In these law codes, data protection also distinguishes between public data and private data, and data sharing takes into account of process problems on collecting, accessing, using, changing, managing, and securing. Specially, the "Bucharest Statement" represents the worldwide mainstream, and believes that the so-called "information society" shall be "based on broad dissemination and sharing of information and genuine participation of all stakeholders – governments, private sector, and civil society".

In regard to computer anti-abuse, the Swedish Data Act of 1973 was the first computer anti-abuse law in the world. Other countries were not far behind. The United States has developed a robust legal system through a series of international conventions, federal laws, state laws, as well as administrative decisions and judicial precedents, including (1) the Computer Fraud and Abuse Act (CFAA), enacted by Congress in 1986; (2) the No Electronic Theft Act (NET), enacted in 1997; (3) the Anticybersquatting Consumer Protection Act (ACPA) of 1999; (4) the Cyber Security Enhancement Act of 2002; and (5) the Convention on Cybercrime (also known as the Budapest Convention) ratified by the United States Senate by unanimous consent in 2006. The Computer Misuse Act (1990) of the United Kingdom declared unauthorized data access, destruction, disclosure, modification of data, and/or denial of service to be illegal. The Council of European formed the Committee of Experts on Crime in Cyber-space in 1997 to undertake negotiations of a draft resolution proposed to an international convention on cyber-crime. Germany amended the 41st Amendment (of the basic law passed in 1994) to the Criminal Code in 2007 against cyber-crimes, including penalties for processing fraudulent transactions, falsifying evidence, tampering materials, and destroying documents. Singapore enacted the Computer Misuse Act in 1993 that was amended in 1998. South Korea established the Critical Information Infrastructure Protection Act in 2001 to implement protective measures against hackers and viruses. Japan enacted an Act on Prohibition of Unauthorized Computer Access in 1999 and made amendments to its Penal Code to expand the scope of criminalization for computer abuse. Australia is a pioneer in establishing data protection principles, including anti-spam legislation, online content regulation, and broadcast services specification. After the famous hacker Aaron Swartz's committed suicide,[1] a wave of widespread skepticism as to whether these laws enacted excessive punishment sparked a debate. Aaron Swartz, aged 26, a well-known computer programmer and Reddit cofounder – but not an MIT student – faced a 35-year prison sentence and a fine of up to US$1 million on federal data-theft charges for illegally downloading, from the MIT computer network, articles from a subscription-based academic database called JSTOR. He pleaded not guilty but hanged himself before trial in his Brooklyn apartment in January 2013. Several prominent observers and Swartz's family criticized the potential penalty for being disproportionate to the alleged crime, claiming "intimidation and prosecutorial overreach" by the criminal justice system to have impelled Swartz to desperation.[2]

Data resources have the general characteristics of natural resources that are used for satisfying our needs. These characteristics include morphological diversity, heterogeneity, and maldistribution. We should reference existing natural resource laws, and ideas for legislation coming from information sharing or from computer anti-abuse, in order to reintroduce data resource laws based on the following viewpoints, instead of mechanically copying from them. First and foremost, we may use references in natural resource law to separate data resource rights into possession, exploration, and development. Second, we may divide data resources into non-/or critical data according to an idea from information sharing legislation: critical data resources should be nationalized with encapsulation of some copies, and the noncritical data resources may be private. Third are the transfer of rights to universities and scientific institutes for exploration and the first-round of development by a bidding process, in order to prevent data resource abuse. Fourth, is to avoid excessive punishment during the data resource development via cascading.

12.2   DATA ANTITRUST LAW

To safeguard a fair competitive market order and facilitate economic development, the major countries with market economies implement their own antitrust laws. Many in the United States have said that the antitrust law is the "Magna Carta of Free Enterprise," whereas in Germany it is part of the "the Economic Constitution."

A monopoly is a structure in which a single supplier (also known as a single seller, a price maker, or a profit maximizer) "produces and sells a given product. Holding a monopoly of a market is often not illegal in itself, however certain categories of behavior can be considered abusive."[3] Such behaviors might be as diverse as capital, technology, or labor, which are manifest in price discrimination, price lock or manipulation, high barriers, exclusive dealing, joint boycotting, and bid rigging. In the data industry, monopolistic performance that directly appears as a data monopoly, involves both a data coercive monopoly and the dictatorship of data [1].

---

[1] http://business.time.com/2013/01/14/mit-orders-review-of-aaron-swartz-suicide-as-soul-searching-begins.

[2] Copied from "official statement from family and partner of Aaron Swartz": http://www.rememberaaronsw.com/statements/family.html.

[3] http://en.wikipedia.org/wiki/Monopoly.

The most prominent example of a data coercive monopoly, of course, is the famous Google's search masked by Facebook. Another known example is that Baiduspider was partially blocked by Taobao.com. Such events euphemistically called "protecting the interests of users," but they are actually monopolistic competitions among large enterprises.

The dictatorship of data, introduced by Viktor Mayer-Schonberger's 2012 book *Big Data: A Revolution That Will Transform How We Live, Work, and Think*, is a government-granted monopoly that features direct intervention in free markets caused by an overreliance on data. The profits from such a monopolistic behavior not only entice enterprises but fascinate government officials who hold the power of administrative examination and approval. This would become a significant threat to the embryonic market order of the data industry, and it should be avoided if possible.

Hence we must recognize the two monopolies as big threats to the balance of competition in free markets, especially in the early stage of the data industry. This gives rise to the need for unified legislations in the field of economic law to establish the new evaluation and transaction mechanisms for data assets and data products, prevent market advantages abuse or excessive government intervention, and to rectify past behaviors that hampered competition or profits. Of course, regulation of monopolies does not mean one opposes scale economics but rather the monopolistic acts themselves.

12.3   DATA FRAUD PREVENTION LAW

The term "authenticity" is used in psychology and existentialist philosophy. It originates from Greek meaning "original" and "self-made" and was introduced to describe "the existence of the real self" in the 1970s. Seeing is believing through our sensory organs, and is a fundamental survival judgment [5] in the real world; however, in cyberspace, seeing leads to confusion with authentic existence, and even has resulted in "identity disorder and self-fragmentation" [84]. According to an article in SFGate.com, Twitter user William Mazeo of Brazil was surprised and angry, when he saw a phony tweet accompanied by his profile picture that said "I wish I could make fancy lattes like in the @barristabar commercial." This data is fraudulent, and is aimed at misleading by confounding right and wrong.

There is a paradox: from a legal viewpoint, determining data authenticity is a premise for building a new mechanism of data fraud; while in turn, identifying the data that are true or false (correct or wrong) requires a judgment mechanism provided by laws and regulations. For example, the Data Quality Act only had one paragraph of twenty-seven words and was enacted by the United States Congress in 2002, to "provide policy and procedural guidance to Federal agencies for ensuring and maximizing the quality, objectivity, utility, and integrity of information (including statistical information) disseminated by Federal agencies." Despite corresponding enforcement guidelines being subsequently issued by relevant Federal agencies, some issues remain unresolved, of which the most critical issue is "who" has "the right of final interpretation" for data quality.

Since "objective truth" cannot be employed in the judgment of data authenticity, we may temporarily try to use "legal truth" for the value judgment of relative truth. Only in this way are we able to apply the provisions given in existing laws to criminal law.

12.4 DATA PRIVACY LAW

Personal privacy protection is undoubtedly the biggest challenge facing the data industry. According to the 1995 EU Data Protection Directive (also known as Directive 95/46/EC), personal data are defined as "any information relating to an identified or identifiable natural person ("data subject"); an identifiable person is one who can be identified, directly or indirectly, in particular by reference to an identification number or to one or more factors specific to his physical, physiological, mental, economic, cultural or social identity," including: natural status, family background, social background, life experiences, and habits and hobbies. Personal data has two significant legal characteristics: (1) the data subject is a "person"; and (2) it "enables direct or indirect identification" of a data subject.

Throughout the world, major personal privacy protection laws can be divided into three categories: (1) comprehensive legislation, represented by the majority of European OECD countries, which have enacted a comprehensive laws to regulate the behaviors of government, commercial organizations, and other institutions in collecting and utilizing personal data; (2) respective legislation, represented by the United States, which uses different laws to respectively regulate; and (3) eclecticism in law, represented by Japan, which has unified legislation as well as different rules and laws for specific domains and sectors.

In 1980, the OECD issued its *Recommendations of the Council Concerning Guidelines Governing the Protection of Privacy and Trans-Border Flows of Personal Data*, and recommended eight principles for the protection of personal data, namely collection limitations, data quality, purpose specification, use limitation, security safeguards, openness, individual participation, and accountability. In recent years, worldwide privacy protection norms have gradually reduced these principles to the right of "whether, how, and who to use" as vested by the data subject. Specifically, there are four rights of a data subject: (1) the right to know, meaning a data subject has a right to know who the data users are, what the data is about, and how the data will be used; (2) the right to choose, meaning a data subject has a right to choose whether or not to provide personal data; (3) the right to control, meaning a data subject has a right to request the data users to use data (e.g., access, disclosure, modification, deletion) in a reasonable manner; and (4) the right to security: meaning a data subject has a right to request that the data users ensure data integrity and security. In real operations the four rights have been formulated as "notice" and "permission." Yet, the PRISM-gate scandal had even simplified the four rights to only one – the right to be informed; with this, the relevant laws become a dead letter.

It is time to revise and expand existing data privacy legislation to make data privacy no longer the "stumbling block" preventing the development of the data industry. We may shift the responsibility of data privacy from a "personal" data subject's permission to data users' shoulders. We recommend four changes that will result in new data privacy norms: (1) a change from deleting all the personal data to removing (or hiding) the "sensitive" private portions only, such as personal identity, religious identity, political preference, criminal records, and sexual orientation; (2) a change from the permanent possession of personal data to possession with an explicit data retention limit (e.g., a time limit may contribute to active transactions in the data markets); (3) a change from using exact match to applying fuzzy data processing; and (4) a change where data mining results are not be applied to data subjects (i.e., we cannot judge whether a data subject is guilty or not in the "future" simply based on potential personal tendencies obtained through behavior pattern mining).

12.5   DATA ASSET LAW

Private law, a part of both the civil law and commercial law, targets longitudinal adjustment of socioeconomic relations, of which civil law stresses the specific form of property and is generally intended to regulate the property and personal relations between equal subjects; commercial law emphasizes the integrity of property and is particularly used to adjust the commercial relationships and behaviors between equal subjects. A data asset, on one hand, is an intangible property that might be in a special form (e.g., electronic securities, virtual currency) and, on the other hand, is a valuable but scarce production materials for data enterprises.

From a development perspective, there are several indispensable steps to enact data asset law: recognize private property rights, clear property rights, and allow property transfer and assignment. First of all, we should clearly recognize that private ownership is vital to data assets. We note the following issues inherent in the application of privacy laws in the adjustment of data assets: (1) Consider the civil law system; data asset law needs to reflect three principles – absolute property, freedom of contract, and fault liability. (2) Consider the commercial code; the data asset should be included as an operating asset that provides substantial value to an enterprise (e.g., patents, copyrights, and trademarks), despite a lack of physical substance; thus transfer and assignment for data assets can be realized. In summary, only when each step has been checked will incentives be provided that directly promote and protect data innovation activities and outcomes, and indirectly change and increase the utility curve of investor behavior, to facilitate the investment and trade in the emerging data industry.

REFERENCES

1. Viktor Mayer-Schonberger (with Kenneth Cukier). *Big Data: A Revolution That Will Transform How We Live, Work, and Think*. Houghton Mifflin Harcourt. 2012.

2. Nicholas Negroponte. *Being Digital*. Vintage. 1996.

3. Raymond B. Cattell. *Intelligence: Its Structure, Growth and Action*. Elsevier Science. 1987.

4. John von Neumann. *The Computer and the Brain*. Yale University Press. 2000.

5. Yangyong Zhu and Yun Xiong. *Dataology (in Chinese)*. Fudan University Press. 2009.

6. Pang-Ning Tan, Michael Steinbach, and Vipin Kumar. *Introduction to Data Mining*. Addison-Wesley. 2005.

7. Martin Hilbert and Priscila López. The world's technological capacity to store, communicate, and compute information. *Science* 2011, **332** (6025): 60–65.

8. Annie Brooking. *Intellectual Capital: Core Asset for the Third Millennium*. Thomson Learning. 1996.

9. Thomas A. Stewart. *Intellectual Capital: The New Wealth of Organizations*. Doubleday Business. 1997.

10. Patrick H. Sullivan. *Profiting from Intellectual Capital: Extracting Value from Innovation*. Wiley. 1998.

11. Max H. Boisot. *Knowledge Assets: Securing Competitive Advantage in the Information Economy*. Oxford University Press. 1999.

12. George J. Stigler. *Memoirs of an Unregulated Economist*. University of Chicago Press. 2003.

13. Tony Fisher. *The Data Asset: How Smart Companies Govern Their Data for Business Success*. Wiley. 2009.

14. Michael E. Porter. *The Competitive Advantage of Nations*. Free Press. 1998.

15. Marc U. Porat. *The Information Economy*. University of Michigan. 1977.

16. Paul M. Romer. Increasing returns and long run growth. *Journal of Political Economy* 1986, **94** (5): 1002–1037.

17. Colin Ware. *Information Visualization: Perception for Design*. Morgan Kaufmann. 2000.

18. Frits H. Post, Gregory M. Nielson, and Georges-Pierre Bonneau. *Data Visualization: The State of the Art*. Springer. 2002.

19. Toby Segaran and Jeff Hammerbacher. *Beautiful Data: The Stories behind Elegant Data Solutions*. O'Reilly Media. 2009.

20. Peter J. Alexander. Product variety and market structure: A new measure and a simple test. *Journal of Economic Behavior and Organization* 1997, **32** (2): 207–214.

21. Karl Marx. *Das Kapital—Capital: Critique of Political Economy*. CreateSpace Independent Publishing Platform. 2012.

22. Dale W. Jorgenson. Information technology and the US economy. *American Economic Review* 2001, **91** (1): 1–32.

23. Tony Hey, Stewart Tansley, and Kristin Tolle. *The Fourth Paradigm: Data-Intensive Scientific Discovery*. Microsoft Research. 2009.

24. Duncan J. Watts. A twenty-first century science. *Nature* 2007: **445–489**.

25. Declan Butler. Web data predict flu. *Nature* 2008, **456** (7220): 287–288.

26. Cukier Kenneth and Viktor Mayer-Schoenberger V. Rise of big data: How it's changing the way we think about the world. *Journal of Foreign Affairs* 2013, **92**: 28.

27. Michele Banko and Eric Brill. Mitigating the paucity-of-data problem: Exploring the effect of training corpus size on classifier performance for natural language processing. In *Proceedings of the First International Conference on Human Language,* pp. 1–5. Association for Computational Linguistics, Stroudsburg, PA, 2001.

28. Tony Hey, Anthony J. G. Hey, and Gyuri Pápay. *The computing universe: a journey through a revolution*. Cambridge University Press, 2014.

29. Raymond Kosala, Hendrik Blockeel. Web Mining Research: A Survey. *ACM SIGKDD Explorations Newsletter* 2000, **2** (1): 1–15.

30. Albert-Laszlo Barabasi. *Bursts: The Hidden Pattern behind Everything We Do, from Your E-mail to Bloody Crusades*. Plume. 2011.

31. Bing Liu. *Web Data Mining: Exploring Hyperlinks, Contents, and Usage Data (Data-Centric Systems and Applications)*. Springer. 2010.

32. David Lazer, Alex Pentland, Lada Adamic, et al. Computational social science. *Science* 2009, **323** (5915): 721–723.

33. John A. Barnes. Class and committees in a Norwegian island parish. *Human Relations* 1954, **7** (1): 39–58.

34. Nicholas A. Christakis and James H. Fowler. *Connected: The Surprising Power of Our Social Networks and How They Shape Our Lives—How Your Friends' Friends' Friends Affect Everything You Feel, Think, and Do*. Back Bay Books. 2011.

35. Francisco S. Roque, Peter B. Jensen, Henriette Schmock, et al. Using electronic patient records to discover disease correlations and stratify patient cohorts. *PLoS Computational Biology* 2011, **7** (8): e1002141.

36. Yun Xiong and Yangyong Zhu. Mining peculiarity groups in day-by-day behavioral datasets. In *Proc. of 9th IEEE International Conference on Data Mining (ICDE 2009)*, 578–587.

37. Jinqrui He. *Rare Category Analysis*. ProQuest, UMI Dissertation Publishing. 2011.

38. Michael E. Porter. *The Competitive Advantage: Creating and Sustaining Superior Performance*. Free Press. 1998.

39. Gary Gereffi, John Humphrey, and Timothy J. Sturgeon. The Governance of Global Value Chains. *Review of international political economy* 2005, **12**(1): 78–104.

40. Arthur Hughes. *Strategic Database Marketing: The Masterplan for Starting and Managing a Profitable, Customer-Based Marketing Program,* 4th ed. McGraw Hill. 2011.

41. Jeremy Ginsberg, Matthew H. Mohebbi, and Rajan S. Patel. Detecting influenza epidemics using search engine query data. *Nature* 2009, **457** (7232): 1012–1014.

42. Renato Dulbecco. A turning point in cancer research: Sequencing the human genome. *Science* 1986, **231**: 1055–1056.

43. Vernon W. Ruttan. *Technology, Growth, and Development: An Induced Innovation Perspective*. Oxford University Press. September 14, 2000.

44. David Shotton, Katie Portwin, Graham Klyne, and Alistair Miles. Adventures in Semantic Publishing: Exemplar Semantic Enhancements of a Research Article. *PLoS Computational Biology* 2009, **5** (4): e1000361.

45. Robert Lipton, Xiaowen Yang, Anthony A. Braga, Jason Goldstick, Manya Newton, and Melissa Rura. The geography of violence, alcohol outlets, and drug arrests in Boston. *American Journal of Public Health* 2013, **103** (4): 657–664.

46. Samuel D. Warren and Louis D. Brandeis. The right to privacy. *Harvard Law Review* 1890: 193–220.

47. Viktor Mayer-Schönberger. *Delete: The Virtue of Forgetting in the Digital Age*. Princeton University Press. 2011.

48. Clara Shih. *The Facebook Era: Tapping Online Social Networks to Market, Sell, and Innovate.* Addison-Wesley. 2010.

49. Stephen A. Ross. The interrelations of finance and economics: Theoretical perspectives. *American Economics Review* 1987, **77** (2): 29–34.

50. Eric Yudelove. *Taoist Yoga and Sexual Energy: Transforming Your Body, Mind, and Spirit*. Llewellyn Worldwide. 2000.

51. William Poundstone. *Priceless: The Myth of Fair Value (and How to Take Advantage of It)*. Hill and Wang. January 2011.

52. Chris Anderson. *The Long Tail: Why the Future of Business Is Selling Less of More*. Hyperion. 2008.

53. Jonathan E. Cook and Alexander L. Wolf. Discovering models of software processes from event-based data. *ACM Transactions on Software Engineering and Methodology* 1998, **7** (3): 215–249.

54. Edward Frazelle. *World-Class Warehousing and Material Handling*. McGraw-Hill. 2002.

55. George B. Dantzig and John H. Ramser. The truck dispatching problem. *Management Science* 1959, **6** (1): 80–91.

56. Peter F. Drucker and Joseph A. Maciariello. *The Daily Drucker*. HarperBusiness. 2004.

57. Paul Timmers. Business models for electronic markets. *Electronic Markets* 1998, **8** (2): 3–8.

58. Alexander Osterwalder, Yves Pigneur, and Christopher L. Tucc. Clarifying business models: Origins, present, and future of the concept. *Communications of the Association for Information Systems* 2005, **16** (1): 1–25.

59. Michael Morris, Minet Schindehutte, and Jeffrey Allen. The entrepreneur's business model: Toward a unified perspective. *Journal of Business Research* 2005, **58** (6): 726–735.

60. Alexander Osterwalder. *The Business Model Ontology—A Proposition in a Design Science Approach.* Institut d'Informatique et Organisation. Lausanne, Switzerland, University of Lausanne, Ecole des Hautes Etudes Commerciales HEC. 2004.

61. Raphael Amit and Christoph Zott. Value creation in eBusiness. *Strategic Management Journal* 2001, **22**: 493–520.

62. Richard Makadok. Toward a synthesis of the resource-based and dynamic-capability views of rent creation. *Strategic Management Journal* 2001, **22** (5): 387–401.

63. Peter F. Drucker. *Innovation and Entrepreneurship.* HarperBusiness. 2006.

64. Eric von Hippel. *The Sources of Innovation.* Oxford University Press. 1988.

65. Clayton M. Christensen. *The Innovator's Dilemma: When New Technologies Cause Great Firms to Fail.* Harvard Business Press. 1997.

66. Henry Chesbrough. Business model innovation: Opportunities and barriers. *Long Range Planning* 2010, **43** (2/3): 354–363.

67. W. Chan Kim and Renee Mauborgne. *Blue Ocean Strategy: How to Create Uncontested Market Space and Make Competition Irrelevant.* Harvard Business Review Press. 2005.

68. Michael E. Porter. *Competitive Advantage of Nations.* Free Press. 1998.

69. Michael Grossman. The demand for health, 30 years later: A very personal retrospective and prospective reflection. *Journal of Health Economics* 2004, **23** (4): 629–636.

70. Masahisa Fujita and Jacques-François Thisse. *Economics of Agglomeration: Cities, Industrial Location, and Globalization.* Cambridge University Press. 2013.

71. John A. Byrne. The virtual corporation. *Business Week* 1993, **8**: 36–41.

72. Constantinos C. Markides. Corporate Refocusing and Economic Performance, 1981–87. Unpublished PhD dissertaion of Harvard Business School. 1990.

73. Alfred D. Chandler Jr. *Scale and Scope: The Dynamics of Industrial Capitalism.* Belknap Press of Harvard University Press. 1990.

74. Michael E. Porter. Clusters and the new economics of competition. *Harvard Business Review* 1998, **76** (6): 77–90.

75. Michael E. Porter. Location, clusters, and the "new" microeconomics of competition. *Business Economics* 1998: 7–13.

76. AnnaLee Saxenian. *Regional Advantage: Culture and Competition in Silicon Valley and Route 128.* Harvard University Press. 1996.

77. Rui Baptista and Peter Swarm. Do firms in cluster innovate more? *Research Policy* 1998, **27**: 525–540.

78. John V. Henderson. Efficiency of resource usage and city size. *Journal of Urban Economics* 1986, **19** (1): 47–70.

79. Edward L. Glaeser. *Triumph of the City: How Our Greatest Invention Makes Us Richer, Smarter, Greener, Healthier, and Happier.* Penguin Books. 2012.

80. Everett M. Rogers. *Diffusion of Innovations.* Free Press. 2003.

81. Theo de Bruijn and Vicki Norberg-Bohm, eds. *Industrial Transformation: Environmental Policy Innovation in the United States and Europe.* MIT Press. 2005.

82. Ernst F. Schumacher. *Small Is Beautiful: Economics as if People Mattered.* Harper Perennial. 2010.

83. Bronwyn H. Hall. *University–Industry Research Partnerships in the United States.* Badia Fiesolana, European University Institute. 2004.

84. Douglas Kellner. *Media Culture: Cultural Studies, Identity and Politics between the Modern and the Post-Modern.* Routledge. 1995.

INDEX